AF805762

History of Mechanism and Machine Science

Volume 22

Series Editor

Marco Ceccarelli, Cassino, Italy

For further volumes:
http://www.springer.com/series/7481

Aims and Scope of the Series

This book series aims to establish a well defined forum for Monographs and Proceedings on the History of Mechanism and Machine Science (MMS). The series publishes works that give an overview of the historical developments, from the earliest times up to and including the recent past, of MMS in all its technical aspects.

This technical approach is an essential characteristic of the series. By discussing technical details and formulations and even reformulating those in terms of modern formalisms the possibility is created not only to track the historical technical developments but also to use past experiences in technical teaching and research today. In order to do so, the emphasis must be on technical aspects rather than a purely historical focus, although the latter has its place too.

Furthermore, the series will consider the republication of out-of-print older works with English translation and comments.

The book series is intended to collect technical views on historical developments of the broad field of MMS in a unique frame that can be seen in its totality as an Encyclopaedia of the History of MMS but with the additional purpose of archiving and teaching the History of MMS. Therefore the book series is intended not only for researchers of the History of Engineering but also for professionals and students who are interested in obtaining a clear perspective of the past for their future technical works. The books will be written in general by engineers but not only for engineers.

Prospective authors and editors can contact the series editor, Prof. M. Ceccarelli, about future publications within the series at:

LARM: Laboratory of Robotics and Mechatronics
DiMSAT – University of Cassino
Via Di Biasio 43, 03043 Cassino (Fr)
Italy
E-mail: ceccarelli@unicas.it

Francis C. Moon

Social Networks in the History of Innovation and Invention

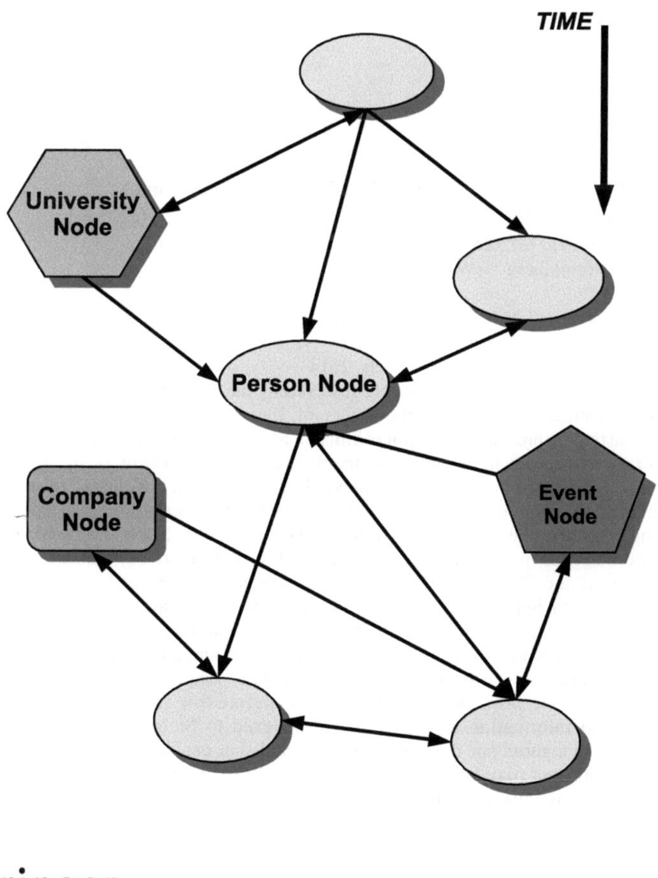

Springer

Francis C. Moon
Sibley School of Mechanical and Aerospace Engineering
Cornell University
Ithaca, NY
USA

ISSN 1875-3442 ISSN 1875-3426 (electronic)
ISBN 978-94-007-7527-5 ISBN 978-94-007-7528-2 (eBook)
DOI 10.1007/978-94-007-7528-2
Springer Dordrecht Heidelberg New York London

Library of Congress Control Number: 2013946318

© Springer Science+Business Media Dordrecht 2014
This work is subject to copyright. All rights are reserved by the Publisher, whether the whole or part of the material is concerned, specifically the rights of translation, reprinting, reuse of illustrations, recitation, broadcasting, reproduction on microfilms or in any other physical way, and transmission or information storage and retrieval, electronic adaptation, computer software, or by similar or dissimilar methodology now known or hereafter developed. Exempted from this legal reservation are brief excerpts in connection with reviews or scholarly analysis or material supplied specifically for the purpose of being entered and executed on a computer system, for exclusive use by the purchaser of the work. Duplication of this publication or parts thereof is permitted only under the provisions of the Copyright Law of the Publisher's location, in its current version, and permission for use must always be obtained from Springer. Permissions for use may be obtained through RightsLink at the Copyright Clearance Center. Violations are liable to prosecution under the respective Copyright Law.
The use of general descriptive names, registered names, trademarks, service marks, etc. in this publication does not imply, even in the absence of a specific statement, that such names are exempt from the relevant protective laws and regulations and therefore free for general use.
While the advice and information in this book are believed to be true and accurate at the date of publication, neither the authors nor the editors nor the publisher can accept any legal responsibility for any errors or omissions that may be made. The publisher makes no warranty, express or implied, with respect to the material contained herein.

Printed on acid-free paper

Springer is part of Springer Science+Business Media (www.springer.com)

This Book is dedicated to the memory of Catherine Elisabeth Waddington [née Moon] 1963–2012

Preface

Like other new ideas, this book had its seeds in earlier works on the history of invention and innovation. My interest began when I was writing about the history of kinematic mechanisms and reading the work of historian Eugene Ferguson. In his work, he traced the invention of many nineteenth century mechanisms back to the 'theater of machines' books of the sixteenth, seventeenth, and eighteenth centuries. Later when I was writing about the great German theoretical engineer, Franz Reuleaux, I began to link up his work to others in England, France, and other countries and was surprised when his name also appeared in the history of the internal combustion engine. It was at a conference in Sicily in 2010 on Archimedes that I felt challenged to address the 'genius theory' of history and decided to posit an alternative model based on technical evolution and networks. I initially addressed the Wright brothers and early aviation, presenting a paper in 2012 in Amsterdam. In this monograph, I examine other inventions and posit a new formalism for addressing the history of technology and applied science.

In presenting the ideas of networks in the context of history, I wanted to outline what I brought from my background as an engineer and to offer a little flavor of modern social network theory without too much mathematical formalism. I have used the barest mathematical notation and have opted instead for graphical and pictorial pedagogy. Also since this book is in a series on the history of machines, I have focused on mechanical and electrical inventions and artifacts, and not on chemical and materials processes. The subject of chaos theory arose naturally when it was clear that its network origins stemmed from the technologies of the internal combustion engine and the electronic circuits of the vacuum tube era.

In a book covering eight technologies and several sciences, I risk being labeled a dilettante. At the same time, most readers will not be familiar with all these fields, let alone the history related to them, and many might find the detailed descriptions of nodes and linkages pedantic and academic. Nonetheless, I view this effort in the spirit of an experimental study and believe that although I wish to advance the big picture, I still have to present the minutia to make the case convincing.

For many of the subjects, I have gone to historical sources when I had access to such material. This was the case with Chaps. 2 and 6. In the former, I had spent 5 years digging into Leonardo's theory of machines and Franz Reuleaux's theory of kinematic mechanisms, the latter on a study leave at the Archive of the Deutsches

Museum in Munich in 2001. In other fields, such as aviation (Chap. 4) and radio (Chap. 5) I used both original sources and comprehensive surveys and biographies. In Chap. 1 and the case study of the steam engine, I used histories from the nineteenth century such as that of Robert Thurston. In the example of chaos theory, I lived its early years from 1978 to 1998. This subject was easy in one sense since I know or knew many of the players but difficult because there was no way I could do justice to all my colleagues in the history of the field.

My only defence for the shallow sweep of some of these fields is that I do not see this book as a history of selected technologies. I posit this work as a different way of viewing history of technology and science in which these dozen areas are simply convenient examples. In all of the examples, there are excellent published histories and biographies and I hope I have mentioned some of the best. However, there is no doubt that I have not surveyed all or even the most important nodes in the networks I have studied. I acknowledge that these networks can never be complete. My only purpose here is to show the complexity with which human societies create new ideas and products and that the innovation networks of the past have characteristics of the social networks of the present.

Aside from the historical thumbprints of the technical areas in this book, I have not embarked on a comprehensive survey of social network theory. I have only mentioned a few concepts and mathematical measures to show that historical networks can be seen in the same way as modern networks. The major difference is the scale of the numbers of nodes; the networks in this book are several orders in size below the number of nodes for Facebook or Twitter, for example. That means we do not have the statistical confidence of having thousands to millions of data points. However, scientists often have to work with small datasets from many physical experiments. It simply means that we cannot make definitive statements regarding the statistics of innovation networks, but we can make inferences and observe trends. If this book inspires a few to take up the task of filling in the blanks I left open, I will deem it a success.

Another difference between this work and modern network research is that I could not data mine citation indices and website usage. My data was gleaned from several hundred books on the history of technology and historical documents to create these historical innovation networks. Modest though these networks are in terms of size, I hope that the book can inspire others with build more complex networks tracing the interconnectedness of science and technology over the last two centuries.

Acknowledgments

The ideas for this book developed in part from a history of science conference in Syracusa Italy, on the "Genius of Archimedes," recently published by Springer in 2010. The author wishes to thank Prof. Teun Koetsier of the Vrije Universiteit, Amsterdam for our lively discussion in Syracusa as well as for his comments on an earlier version of Chap. 4 of this book.

Preliminary copies of the manuscript were sent out for review, and the author wishes to thank the following for their helpful comments, suggestions, and encouragement: Prof. David Ollis, North Carolina State University; Prof. Steven Danyluk, Georgia Tech; Prof. Teun Koetsier, Vrije University, Amsterdam; Prof. Robert Friedel University of Maryland; Terry S. Reynolds, Emeritus Professor of History, Michigan Technological University, Prof. Philip Holmes, Princeton University. At Cornell University, thanks are due to Elaine Engst University Archivist, Richard Rand and Steven Strogatz, Mathematics, and Ronald Klein, Science and Technology Studies. The author also thanks aviation historians Richard Hallion and Anne Millbrooke for their comments and suggestions.

Archival material was accessed at the following institutions; Cornell University Library, Rare Books, and Manuscripts (Willis Carrier material, Alfred Zahn, and Glenn Curtiss papers); Columbia University Rare Books Division (Edwin Armstrong Papers); Deutsches Museum Archives, Munich (Franz Reuleaux papers, Lilienthal drawings); Biblioteca Ambrosiana, Milan (Francesco di Georgio manuscript); Notre Dame Archives (Alfred Zahm papers).

The author also found visits to the following museums helpful in writing this book: Deutsches Museum, Munich (Reuleaux kinematic mechanisms, Otto-Langen engines); Wright Brothers Museum, Kitty Hawk, NC; Glenn Curtis Aviation Museum, Hammondsport, NY; Royal Observatory Greenwich, England (Harrison Clock Collection); Science Museum London (Steam Engines), Cornell University Reuleaux Kinematic Models Collection.

Thanks are also due to Professor Marco Ceccarelli, University of Casino, Italy, the Editor of this Springer Series for his encouragement and tireless leadership of the History of Mechanism and Machines (HMM) Commission of the International Federation for the Promotion of Mechanism and Machine Science (IFToMM). The HMM Workshops and Symposia of IFToMM in Cassino, Dresden, Moscow, Ithaca, Taiwan, Amsterdam in the last decade have been especially helpful to the Author in writing this book.

Thanks also go to the staff at Springer Dordrecht, especially Nathalie Jacobs, Senior Publishing Editor for Engineering, and Johanna F. A. (Anneke) Pot, Editorial Assistant. The Author also thanks former student Jocelyn Kluger, now at MIT, for computer assistance.

Special thanks are due to my wife of 50 years Elizabeth Moon who helped to edit the manuscript and to my daughter Catherine (Moon) Waddington who before she passed away last year, sent me and brought to my attention Benoit Mandelbrot's autobiographic history of fractals.

Contents

Chapter 1
Introduction: James Watt's Social Network

Abstract We introduce ideas from network and graph theory and apply them to the evolution and network of contributions of the steam engine. Also discussed are models of growth of social networks, exponential growth and the 'S' curve. This chapter also reviews earlier work on evolution, diffusion and networks of innovation.

Did James Watt invent the steam engine? Did the Wright Brothers invent the airplane? In this book we examine the interactions between the genius-inventor and his or her social network using new ideas centered on modern networking systems. We also present evidence that invention is not just a moment of epiphany in a lone genius inventor, but is a culmination of an evolutionary process resulting from a network of people and institutions. In this spirit, our focus is primarily on the innovation network per se and not on the specific personalities for which there are many books and references. In this endeavor we take an experimental point of view by using the data from the historical record to explore models for innovation and invention.

Modeling innovation history using scientific constructs to uncover universal patterns has precedents in related fields such as biophysics and bioengineering. Many human systems such as epidemiology, traffic flow dynamics and demographics are the subject of mathematical modeling. Why not use network theory to uncover patterns in creative activities such as invention and innovation?

In the last two decades new information technologies have enabled millions of people to engage in social networking on a scale that was unimagined decades earlier. A byproduct of the growth of the World Wide Web, Facebook and Twitter has been a set of research studies on social networks resulting in popular books by Watts (2003), Barabasi (2003) and others. These studies suggest that we are in a new age of human communication and social dynamics that can stimulate innovation as well as trigger political change in places such as the Middle East and Asia through the use of social networking technology. Such claims prompt one to ask if communication networks played a role in earlier human activities in innovation and invention.

This author was first inspired to explore the role of networks in early theories of machine design beginning with the age of Leonardo da Vinci and spanning the industrial revolution in the nineteenth century (Moon 2007). A second study examined social networking in early aviation; did the Wright really invent heavier-than-air aircraft? (Moon 2012) The results of these earlier studies led the author to examine

F. C. Moon, *Social Networks in the History of Innovation and Invention*,
History of Mechanism and Machine Science 22, DOI: 10.1007/978-94-007-7528-2_1,

© Springer Science+Business Media Dordrecht 2014

other technical and scientific innovations such as maritime clocks, internal combustion engines, early radio and more recently chaos theory.

Using models to understand the history of innovation has roots in Everett Rogers' *Diffusion of Innovation*, first published in 1962 and lately reissued. More recently there have been several books that describe the application of complexity theory to the study of history e.g. Mark Buchanan's *Ubiquity* (Buchanan 2000). Buchanan describes a new model called *self-organized criticality* or SOC to encompass several complex phenomena such as earthquakes and the growth of cities. Some of these theories focus on the propagation of consumer products into the marketplace made popular in a contemporary book *The Tipping Point,* by Gladwell (2000). Understanding market diffusion of new products and how they stimulate other products, as in the case of Apple's iPhone and the thousands of so-called 'Apps', is important. But the broader question is how the basic scientific and technical ideas behind these products evolved. In this book we focus on the early creative processes and the networks that produced these new ideas and products.

Beginning in the 1950s, physicist and historian of science Derek de Solla Price of Yale, brought mathematical methods to the sociology of science (1951). He showed that the growth of research publications of the past century followed an exponential function, doubling in 10–15 years. In another work, de Solla Price (1965a, b) studied the network of scientists by analyzing the citations they made of each others research papers. De Solla Price used statistical techniques to measure the productivity of scientists. In this book we seek a more detailed analysis of technical innovation by following the networks of individual contributors. We hope to provide some evidence for the statistical patterns of scientific and technical history through the action of the growth of networks.

Using graphics to encapsulate events and historical patterns has a long history as illustrated in the recent book *Cartographies of Time; A History of the Timeline*, by Rosenberg and Grafton (2010). They trace the use of chronologies in history to the Greeks and later to the fifth century CE. At the time of the steam engine, the British scientist, Joseph Preistley, a friend of Benjamin Franklin, published a *Chart of Biography* in 1765 with names of historical figures and dates of birth. We use event timetables in our monograph to search for exponential growth in several examples of innovation and invention.

A more recent example of using networks in historical studies is Randall Collins' *The Sociology of Philosophies* (1998) in which he analyzes the great philosophical traditions from the Greeks to the Chinese using graphical networks to trace the influence of philosophers across generations. He calls his work a history of intellectual networks and makes a case for the proposition that innovative ideas are found in groups and not in the isolated thinker.

Recently young historians, many from Germany and Austria, have formed an on-line network for the study of "Historical Network Research". [http://www.historicalnetworkanalysis.org] Areas of research include applications of network analysis to archeology, history of guild systems in the Holy Roman Empire, history of the Byzantine empire [see Preiser-Kapellier in the Bibliography], sociology of science and more. A broader organization is INSNA: International Network for

Social Network Analysis. Increasingly historians are using computer and quantitative methods to address questions in history and this monograph is hopefully a positive contribution to the use of networks in history of science and technology.

1.1 The Genius Theory of Innovation

As a society we are reluctant to ascribe all human behavior to the constraints of mathematical laws. We still revel in the myth of human geniuses: Archimedes, Newton, Madame Curie, Edison, Einstein and the Wright brothers. More recently we have elevated the late Steven Jobs of Apple into this pantheon. Stories of genius, trial and perseverance inspire and thrill us, especially the young and impressionable, in the vicarious process of human creativity,. However entrenched the genius-hero model of innovation is in modern society, there has long been recognition among historians of science and technology of the evolutionary nature of invention as well as the social connectedness of scientific and engineering achievement. For example the historian Usher (1929) write about the history of machines;

"*The more serious recent literature on the theory of invention theory recognizes the necessity of presenting invention as a cumulative synthesis of a relatively large number of individual items*."

In the nineteenth century, Thurston (1878) in a history of the steam engine argued,

"*Great Inventions are never, and great discoveries are seldom the work of any one mind*."

Perhaps inspired by the emergence of the *Theory of Evolution*, nineteenth century historians began to imagine that invention was also part of such a process, though on a shorter time scale.

A modification of the genius theory of innovation incorporates an evolution of scientific and technical ideas through the vehicle of successive genius-inventors. This idea is sometimes captured by the phrase '*on the shoulders of giants*" and was the title of a popular book. This theory is represented in Fig. 1.1 where ideas of human-powered flight begin with Leonardo and progress to the Wright Brothers without acknowledging the contributions of dozens if not hundreds of others. A more complete picture is presented in Chap. 4.

In physics we can draw a similar evolutionary genius chart beginning with Galileo and Newton in classical physics and ending with J. Clerk Maxwell and Einstein and Special Relativity, neglecting many major and minor contributors and institutions such as the Royal Society of London or parallel French and German scientific societies. In Chap. 6 of this book we trace the genius theories of nonlinear dynamics of chaos that became popular in the 1980s and use network analysis to illustrate the complexities of scientific progress.

Our plan for this book is to take historical evidence and construct network diagrams for eight different technologies and related sciences. We will also use historical data to plot innovation growth curves to seek evidence for innovation avalanches.

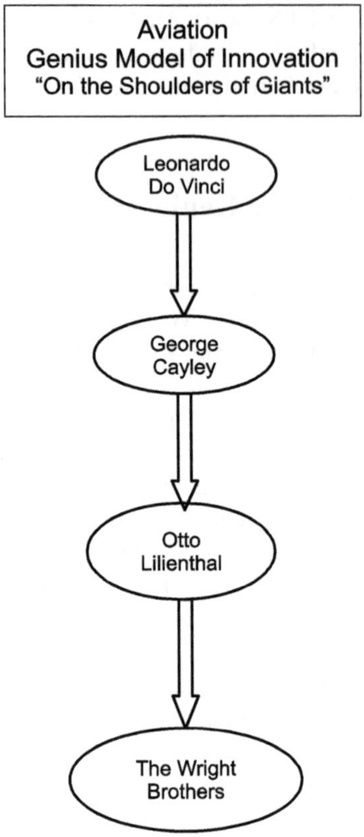

Fig. 1.1 An evolution chart for the genius theory of aviation

It is our premise that the exponential growth in technical innovation is linked to the growth of innovation networks.

1.1.1 Goal of this Monograph

The goal of this monograph is to provide further evidence to challenge the 'Genius Model' of innovation and to make a case that earlier inventions grew out of social networks with characteristics similar to that of the World Wide Web and other modern communication networks. In this chapter we examine the evolution and network of contributions to the steam engine. In Chap. 2 we study the kinematics theory of machine design from the fifteenth to the nineteenth centuries beginning with one of the prime heroes of the genius theory, Leonardo da Vinci. As a particular example of complex machines we construct a network chart for the evolution of precision marine clocks. In Chap. 3 we describe the network of contributors to the creation of the internal combustion engine and its effects on the automotive industry focusing on Henry Ford.

This discussion of engines naturally leads to early aviation in Chap. 4, where we describe a complex web of over 80 individuals and institutions covering the period of 1810–1910 from the time of George Cayley to the Wright brothers and Glenn Curtiss. This case study will provide examples of the importance of 'weak links' and the role of what Gladwell (2000) calls 'Connectors' such as Octave Chanute. We also describe in this chapter the use of an influence matrix and statistics of a link-node distribution curve.

The early development of wireless and radio communication is our fifth case study in Chap. 5. The seeds of wireless telegraphy grew out of the network surrounding the magnetic telegraph. Here again in addition to Marconi, Deforest and Armstrong, there is a complex set of up to fifty players spanning the work from Maxwell [c.1865] to the early days of radio and RCA's David Sarnoff in the 1920s.

It is interesting that inventors who show up in one technology network were also active in one or two other technologies. For example Lenoir, one of the pioneers of the internal combustion engine, received an award from France for his invention of an automated telegraph in 1865. Daimler, whose name is closely identified with automobiles and internal combustion engines, designed a lightweight engine for Count Zeppelin's dirigibles. The Rumkorff induction coil used in early internal combustion ignition systems was also employed in early wireless telegraphy systems. In the mid to late nineteenth century, one invention led to another, one scientific discovery was applied to several new inventions. Thus it should not be a surprise that social networks for different innovations often were interlocking through either common person or institution nodes or common critical technologies.

The mathematical and scientific ideas under the name of *Chaos Theory* are a branch of the mathematics of *nonlinear dynamical systems* In Chap. 6 we show how the seeds of twentieth century Chaos Theory in mathematical physics began to grow in the networks associated with the internal combustion engine and radio electronics. In our sixth detailed study of innovation we trace the mathematical discoveries of Henri Poincare in France at the beginning of the twentieth century to the weather models of Edward Lorenz at MIT and electrical circuits of Yoshisuke Ueda in Kyoto a half-century later. Chaos Theory has since morphed into *ComplexityTheory* in the late 1990s that in turn led to social network theories in the last decade. This example was chosen because of the author's close familiarity with chaotic and nonlinear dynamics, (Moon 1992), and because some fragments of the history of this science have begun to emerge in recent years.

In the Summary, Chap. 7, evidence for other historical innovation networks in science and technology, such as involved George Carrier in the development of air conditioning, will be presented though not in as detailed format.

In this final chapter we gather our data on the exponential rise of technical networks across two centuries, and present evidence for the rapid decrease in doubling time doubling time for growth of new technologies in the twenty-first century. Innovation networks that took half a century to grow in the late eighteenth century now have a doubling time of a year or less. We hope to bolster the thesis that for at least the last 200 years, human society has progressed through the tension of both global cooperation and competition in the evolution of new technologies though social networks of people and institutions.

1.2 James Watt and the Steam Engine: Genius or Social Network?

As an introduction to the ideas in this book we examine the classic case of the evolution of steam power and describe the use of evolution diagrams, network diagrams and influence matrices. Our period of interest spans 1700–1800 before the exponential rise in the use of steam engines for manufacturing and transportation in the nineteenth century. This case is of interest because it predates the birth of the first communications revolution of the magnetic telegraph and the construction of rail networks. Our reference for the nodes and links in the steam engine network are classic histories by Thurston (1878), Lardner (1836) and Farey (1827). From Thurston we learn that preliminary engine trial and error was conducted in the seventeenth century. Here Huygens had proposed a gunpowder engine. Thomas Savery in England began to tackle the pumping of water out of deep mines in Cornwall. Papin an assistant to Huygens in Paris, went to England where he met Robert Boyle and introduced the moving piston into the steam engine. Thomas Newcomen corresponded with Robert Hooke and likely knew of Savery's engine before taking out a patent with John Calley on the first commercial steam engine for deep mines. Thus before the first steam engines were employed in the early eighteenth century, there was already established a network of scientists and inventors.

Principal Nodes in the Steam Engine Network 1700–1800:
Thomas Savery [1650–1715] built a working engine without a piston.
Denis Papin [1647–1712] Introduced a piston into the steam engine.
Thomas Newcomen [1664–1729] Built first partial vacuum, piston steam engine, 1712 for deep mine water pumping.
Joseph Black [1628–1799] Professor at Glascow University; studied latent heat of steam that led to James Watt's condensor.
James Watt [1736–1819] Instrument maker at Glascow University; added a condensor and other improvements to the Newcomen engine that significantly raised efficiency. Basic patents, circa 1769–1785.
Matthew Boulton [1728–1809] Business partner of James Watt. Used his factory system to produce Watt type engines.
John Smeaton [1724–1792] Early builder of Newcomen engines.
Jonathon Hornblower [1753–1815] In 1781 he invented a two piston, compound engine with low and high pressure cylinders, later built by Woolf.
Richard Trevithick [1771–1833] Built one of the first high-pressure steam engines and applied it to a moving vehicle.

Oliver Evans [1755–1819] American inventor who built a high-pressure steam engine, circa 1800.

1.3 Innovation Evolution Models and Network Diagrams

Historians of technology have long described models based on the evolution of innovation and inventions. (e.g. Basalla 1988) The economist and historian Joel Mokyr of Northwestern University has written a book, *The Lever of Riches* (Mokyr 1990), describing technological evolution from antiquity to the nineteenth century. He uses evolutionary concepts of selection and mutation to examine its impact on economic dynamics.

Sometimes innovation evolution is presented in a flow chart such as the one in Fig. 1.2 representing the succession of technologies leading to steam power. In this model, the nature of the connection between each technology is left vague.

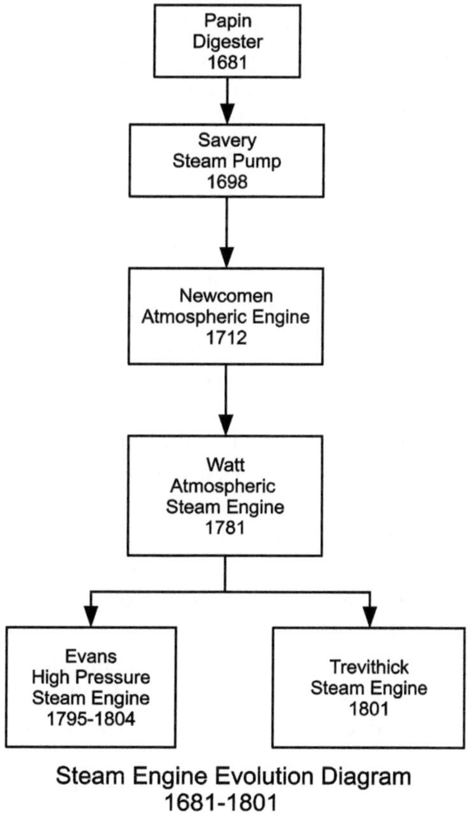

Fig. 1.2 Technology evolution flow diagram for the steam engine. Links do not denote any specific connection between inventions

To construct a social network diagram of steam engines we consult the work of Robert Thurston, [1839–1903], an American academic engineer with considerable practical experience with steam power. Thurston first taught at Stevens Institute of Technology in Hoboken, New Jersey in 1871. In 1873 he was appointed ambassador to the International Exhibition in Vienna. In 1885 the President of Cornell University persuaded Thurston to come to Ithaca and reorganize the College of Mechanical Engineering. Thurston's interest in machine design was on materials and thermodynamics, especially as they impacted the steam engine. Thurston was a firm believer in the evolution of technology. In his well-known treatise, *A History of the Growth of the Steam Engine,* Thurston (1878) wrote:

"*I propose to call attention to the fact that the history [of the steam engine] illustrates the very important truth: it Great inventions are never, and great discoveries are seldom, the work of any one mind. Every great invention is really either the aggregation of minor inventions or the final step of a progression. It is not a creation, but a growth—as truly as is that of the trees of the forest.*"

In his *'History'* Thurston recited a litany of earlier contributors and inventors who made the steam engine possible. He began with the ancient Greeks—Hero's *aeolipile*, a rotating sphere with two arms expelling steam. He recognized the contributors in the Renaissance; Leonardo for his 'steam cannon' or *architonnerre*, and a Spaniard named Blasco de Garay [c. 1543], the Italian Giovanni Battista della Porta [c. 1601], and the French machine book author Solomon de Caus [1615]. From England Thurston named Edward Somerset, Marquis of Worcester, [c. 1663], and from the Netherlands Christian Huygens [c. 1680] who proposed a gunpowder engine as well as the Englishman Sir Robert Moray, Master Mechanic to the King, who measured the pressure-volume properties of steam.

Thurston's litany of working steam engines began with Thomas Savery, [c. 1698], Denys Papin [c. 1687], and finally the blacksmith from Dartmouth England, Thomas Newcomen [c.1705] whose machine concept had a life of over 75 years before the major contributions of James Watt (Fig. 1.3).

These new practical machines were considerably more complex in the variety of machine elements and kinematic mechanisms than the simple cylinder and piston of Leonardo da Vinci or the spherical vessel and cock valve of Solomon de Caus. The steam engines of Newcomen, Boulton and Watt and later engineers employed slider-crank and eccentric mechanisms, planet-sun epicycloid gear trains, straight-line linkages, flywheel and rotating ball speed regulators as well as complex valve control linkages (Fig. 1.4). This complexity increased further when the steam engine was employed in locomotives for railroad transportation. The evolution and network links for kinematic mechanisms in machines are discussed in Chap. 2.

The many variations of the steam engine in the nineteenth century spawned technical reviews and historical books. Two early histories were those by Farey (1827) and Lardner (1836). Thurston's *History of the Growth of the Steam Engine (Thurston 1878)* has become a classic and we use this source to construct a historical social network diagram for the steam engine as shown in Fig. 1.5

Fig. 1.3 Portrait of James Watt from the national portrait gallery London. Watt is studying his famous straight-line mechanism for converting the rotary motion of the rocker arm into straight-line motion of the piston. [Copy of this painting at Cornell University, School of Mechanical and Aerospace Engineering.] See also Fig. 1.4

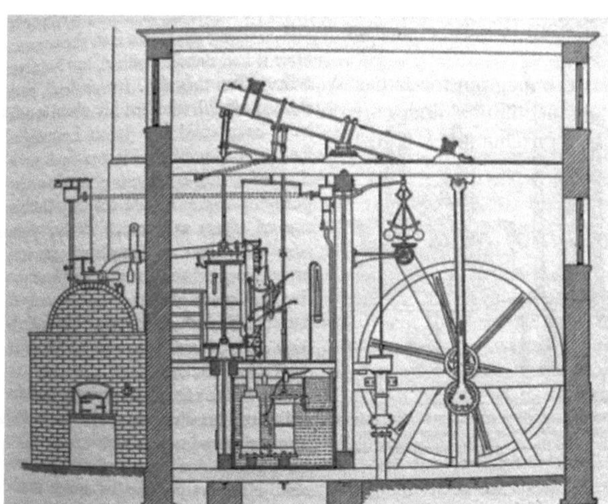

Fig. 1.4 Sketch of steam engine components of James Watt, 1784

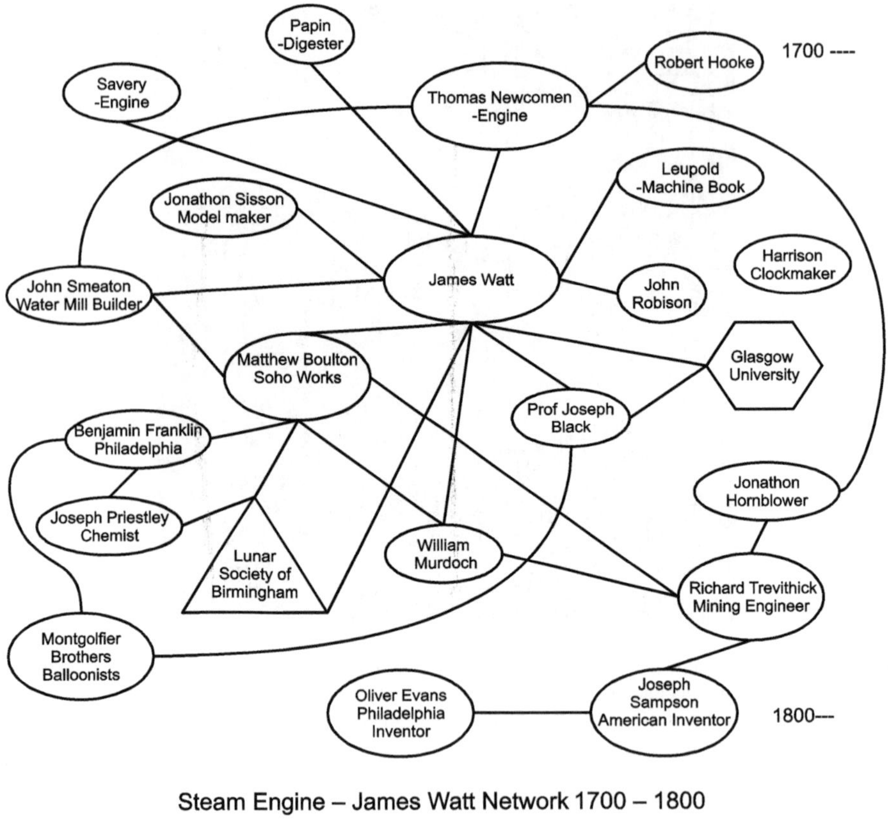

Fig. 1.5 Social network diagram for James Watt and the evolution of the steam engine. Human nodes are ellipses and institutional nodes are polygonal figures. Links denote specific interactions between nodes. [Based on Thurston (1878).]

1.3.1 Description of Selected Links in the Steam Engine Network

The many books on James Watt's contributions to the steam engine (see e.g., Rosen 2010), often tell of Watt's moment of epiphany on Glasgow Green when he conceived of the idea of a condenser for Newcomen's engine. But a closer look will reveal a human network surrounding the inventions of James Watt and his partner Matthew Boulton, connections between inventors, scientists and institutions (Tann 1981) (Fig. 1.5). Links between nodes include personal contacts such as developed in the Lunar Society of Birmingham as well as letters, books and historical links. This network does not diminish the contributions of Watt to steam power, but makes more explicit his connections to a community of other contributors to the technology and to the science of steam power and thermodynamics.

This network in Fig. 1.5 shows the connection between artisan inventors such as James Watt and scientist Professor Joseph Black of Glasgow University. It also shows the connections between Newcomen's earlier steam engine and engineers Smeaton and Trevithick who tried to improve the Newcomen engines without violating the Boulton-Watt patents. Although this incomplete network only contains 22 nodes, it also shows the connection between Great Britain and North America through the links of Benjamin Franklin and Oliver Evans [1755–1819]. Evans received America's third patent and invented a high-pressure steam engine that eventually replaced the Watt atmospheric steam engine. The Lunar Society node is based on a book describing the network of men in and near Birmingham who met regularly and exchanged scientific and technical ideas (see Uglow 2002).

After reading Thurston's history of steam power and thermodynamics and reflecting on the network of Fig. 1.5, can we really say who invented the steam engine? As we shall see in the other examples in this book ascribing credit for new inventions and innovations to one or two individuals is almost impossible in the light of the evidence of a more detailed history of invention. Yet it is difficult for our society to let go of the appellation "inventor".

1.3.2 James Watt's 'Epiphany' Story

One important characteristic of the genius inventor myths is the moment-of-truth or epiphany story, often told many years later by the inventor and subsequently embellished by personal friends, family or corporate interests. Robert Thurston's history of the steam engine quotes Watt's narrative of his ah-ha moment;

"*I had gone for a walk on a fine Sabbath afternoon—{on Glasgow commons}—I was thinking about the engine at the time and had gone as far as the herd's house when the idea come into my mind that, as steam is an elastic body, it would rush into a vacuum and if communication there made between cylinder and an exhausted vessel, it would rush into it and might be there condensed without cooling the cylinder."—" I had not walked further that the Golf house when the whole thing was arranged in my mind.*"

It is not difficult to see how such a story would appeal to a biographer as well as his or her readers: the instant revelation, the magic moment of creativity. Another interpretation of these innovation stories is the 'avalanche' model where the historical network of the information commons becomes sufficiently complete that someone in the network is able to piece the puzzle together and create a new idea or product.

1.3.3 Links to the Aeronautics Network

The network in Fig. 1.5 also shows links between inventors and engineers of steam power and of the nascent field of aerial flight and ballooning. Franklin was a witness to the flights of the Montgolfier brothers in 1783 in Paris, and sent a letter to Joseph Banks at the Royal Society that led Boulton and Priestley to generate hydrogen gas and demonstrate ballooning principles in England. As we shall see in Chap. 4 on early aviation, George Cayley began to ponder the difficulties of using steam engines in aerial machines and speculated on the use of internal combustion engines for flight vehicles.

1.4 History of Innovation Studies

There are many definitions of innovation. In this book, we use the term '*innovation*' to denote new scientific and technical *ideas* and the term '*invention*' to denote new *artifacts* and *products* such as the steam engine, radio or cell phones, for example.

Scholars have posited the idea of a societal role in innovation and evolution. The idea of *prosopography* in sociology was developed in the 1930s as the study of the interaction of a number of players in history. It also has roots in the nineteenth century through Francis Galton [1822–1911] who used many biographies of British scientists to study heredity, environment and genius. The concept of collective biographies was used by the German historian Theodor Mummisen to the study of the Roman Empire. He defined;

> Prosopography [as] a modern word for the study of individual persons in a larger context—[it] aims to establish the social context of groups such as their ethnic and regional origins, family connections and careers (see Keats-Rohan 2007).

Earlier studies such as Everett Rogers (1962) classic book examined the diffusion of innovations such as agricultural science and medical practice in rural America. Rogers used the term 'social systems' instead of 'social networks' and included in his system individuals and organizations. He also referred to communication channels and used the 'S' curve, discussed below, to model the acceleration and decline of innovation into a marketplace. Later studies have examined the penetration of new technologies into society. For example, Hughes (1983) has studied the spread of electrification in Western societies in his book *Networks of Power.*

Although there are precedents in prosopography and network theory connecting different persons in the evolution of technical and scientific ideas, the *Historical Innovation Networks* discussed in this book extends the ideas of prosopography by using elementary, mathematical constructs of network theory.

The use of evolution and network diagrams has been used in the history of art. In 1936, Alfred Barr, the first director of the Museum of Modern Art (MOMA) in New York, drew a famous evolution diagram connecting cubism and abstract art. In a recent exhibition at MOMA, "*Inventing Abstraction*" Paul Ingram of Columbia University presented a network diagram connecting many of the artists in the period 1900–1930 who were creating abstract art (Dickerman 2013).

The Lithuanian-American artist George Maciunas, a member of the New York avant-garde art scene in the 1960s, developed a 6-foot by 12-foot long graphical history of art. A recent book, *Maciunas' Learning Machines,* by Schmidt-Burkhardt (2011) describes the extensive use of complex network diagrams by Maciunas connecting many of the players in the Dada and avant-garde art movements of the early twentieth century.

In the modern era, mathematicians have constructed family trees tracing the students of famous and not so famous professors of mathematics. [See the Mathematics Genealogy Project of the American Mathematics Society] Using the Internet, we can also trace the network of citations of scientific papers connecting scientists in one field with others in disparate fields who cite the same paper.

Science Network Studies

In an earlier work, de Solla Price (1965a, b) studied the network of scientists by statistically analyzing the citations they made of other research papers. He showed that the number of papers N that received k citations N(k) followed a decreasing power law, that is a few papers were cited frequently while most were hardly cited at all. He also showed that as a scientific field evolves, at least half the citations are to papers recently published. This implies that the leaders of a field mainly cited the their current peers and that there was a tight group of scientists in any developing field who were in constant awareness of each other's work. Science progresses in a wave with a tight knit group of researchers at the forefront.

Since the work of de Solla Price, there has been a continuing study of the sociology of science sometimes called '*scientometrics*'. Many of these studies rely on the research citation data readily available on the Web. Citations may be considered 'formal' links in a social network model but they do not account for the informal links such as letters and email exchanges, or oral exchanges, international meetings and workshops and personal contacts. The studies presented in this book rely on data in many narrative histories of a specific technical field. Hence the size of the data sets of nodes and links is less than hundreds and much less than the thousands of citations in the study of contemporary science. Nonetheless, the emerging patterns of exponential growth of nodes and links and the concomitant growth in machine performance and market penetration suggest that technological networks behave in ways similar to that of science but with more subtle kinds of links.

Finally in the fields of electrical engineering and computer science there is a mathematical theory of networks that uses graph theory and matrix mathematics to represent the connections between elements or nodes in the network (see e.g., Newman 2003). In our book, we sample each of these traditions of innovation and network theory and attempt to apply some of these ideas to examples from the history of technical and scientific innovation, especially from the late eighteenth to the early twentieth centuries.

1.5 Historical Innovation Networks

A *historical innovation network* consists of a set of persons [inventors, engineers, scientists, artisans], organizations [e.g. universities, Royal Societies] and social events such as international expositions, each of which is called a *node*. This collection of nodes has inter-nodal connections called *links* (or edges). A general historical innovation network is shown in Fig. 1.6. Links between person-nodes are created

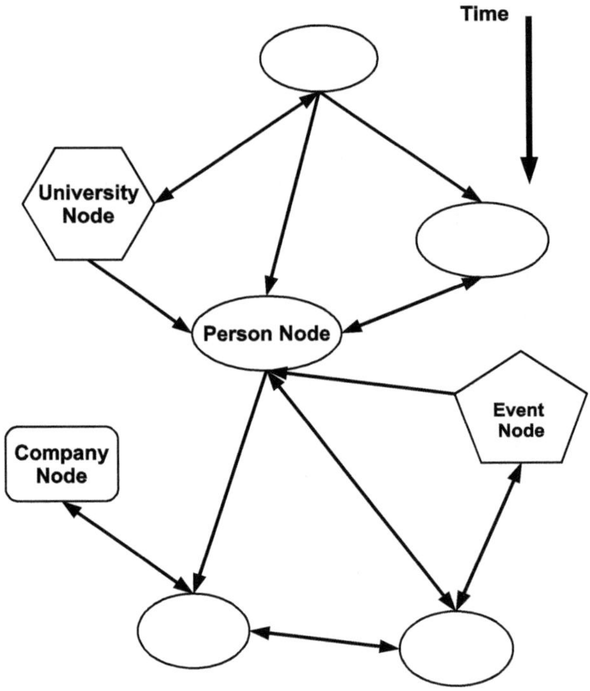

Fig. 1.6 General components of a historical innovation network diagram

by friendship, letter exchange, writings, books, and financial interest, to mention a few. Links between person-nodes and organizations involve membership, professorial appointments, education of students etc. Connections between event nodes and person-nodes are created by attendance or by proceedings etc.

Each link is either directional or bidirectional. Links between nodes in different generations are unidirectional. For example in early aviation Lilienthal could influence the Wright brothers but they did not influence Lilienthal because he died in 1896 three years before the brothers became interested in flight [1899]. In social network theory, nodes can have attributes called *fitness*, which measures the importance of the node or link. For example in aviation, the Wright brothers and Lilienthal would be high fitness nodes. In the internal combustion network, Nikolas Otto would be a high fitness node. Unlike social networks such as the World Wide Web, or *Facebook*, a history network spans generations linking person nodes living in one generation with dead person nodes in an earlier generation. In a history network, nodes can be connected across generations. Historical Innovation/Invention Networks [HIN] also evolve as new nodes are added and new links are created between new players. In HIN, there can be *sub-networks* that can focus on a specific component; e.g. in aviation one can focus on lightweight engine evolution, or the wing design as well as the underlying aerodynamic theory development. In radio one can focus on wireless telegraphy or on vacuum tube development.

The difference between a pure narrative history and a history network theory, is that in the latter, the historian is obliged to determine if there are links between all of the nodes in the network. Thus if a new node (player) is mentioned, its links to all the other nodes must be established. The analysis in this book uses an influence matrix, a network graph and an event-node timeline graph.

1.5.1 Influence Matrices

An *influence matrix* is an array of numbers M[NS, NT] where NS are the source nodes of the influence (rows) and NT are the terminus nodes of the directed influence links (columns). For example in the eighteenth century steam engine network, Newcomen is an NS link for James Watt, whereas Watt is a source node for the Trevithick, the inventor of a steam railroad engine. The matrix for the steam engine network in Fig. 1.7 is 21×21 nodes and the entries are not weighted and contain only zeroes and ones. In this book, node-pairs that were contemporary were assigned both outgoing and incoming links; it was assumed that each influenced the other. Earlier generation nodes connecting later nodes were assumed to have only outgoing links. A more detailed HIN would include weighted, as well as directed links. We could also scale the importance of the nodes with respect to the advance of the technology, however we did not attempt to assign a such node scale for the case studies in this book.

A related representation is the *Adjacency Matrix* for the steam engine network shown in Fig. 1.7. Without assigning direction of influence, this matrix is symmetric and only the lower half is shown. The nodes were listed arbitrarily by numbers of links attached to the node. Not surprisingly, James Watt and Michael Boulton had the

Node	1	2	3	4	5	6	7	8	9	10	11	12	13	14	15	16	17	18	19	20	21
1 Watt	0																				
2 Boulton	1	0																			
3 Newcomen	1	0	0																		
4 Trevithick	0	1	0	0																	
5 Smeaton	1	1	1	0	0																
6 Black	1	0	0	0	0	0															
7 Murdock	1	1	0	1	0	0	0														
8 Lunar	1	1	0	0	0	0	0	0													
9 Franklin	0	1	0	0	0	0	0	0	0												
10 Priestly	0	0	0	0	0	0	1	1	0	0											
11 Glasgow	0	0	0	0	0	1	0	0	0	0	0										
12 Hornblow.	1	0	1	1	0	0	0	0	0	0	0	0									
13 Montgolfi.	0	0	0	0	0	1	0	1	0	0	0	0	0								
14 Sampson	0	0	0	1	0	0	0	0	0	0	0	0	0	0							
15 Evans	0	0	0	0	0	0	0	0	0	0	0	0	0	1	0						
16 Robison	1	0	0	0	0	0	0	0	0	0	0	0	0	0	0	0					
17 Leupold	1	0	0	0	0	0	0	0	0	0	0	0	0	0	0	0	0				
18 Sisson	1	0	0	0	0	0	0	0	0	0	0	0	0	0	0	0	0	0			
19 Savery	1	0	0	0	0	0	0	0	0	0	0	0	0	0	0	0	0	0	0		
20 Papin	1	0	0	0	0	0	0	0	0	0	0	0	0	0	0	0	0	0	0	0	
21 Hooke	0	0	1	0	0	0	0	0	0	0	0	0	0	0	0	0	0	0	0	0	0

Fig. 1.7 Adjacency matrix for the influence matrix of James Watt and the steam engine social network based on Fig. 1.5 and Thurston (1878)

highest number of links. This suggests a modified definition of the genius theory of invention where the nodes with the greatest influence or number of links are anointed one of the principal innovators. This approach recognizes the importance of creative people as well as acknowledges the role of other players in the innovation network.

1.6 Introduction to the Theory of Graphs

The above example is one of many similar problems that fall under the mathematical study of *graph theory* (see e.g. Wilson and Watkins 1990). Such problems date back to the time of Euler in the eighteenth C. Graphs arise in such diverse areas as computer science, electrical circuits, transportation networks, economics and in biochemistry of macromolecules. The National Institute of Health in the US has a "Human Connectome Project" run by UCLA. Using brain scans scientists can trace the neural pathways through the brain.

Recently scientists have used graph theory to address the complexities of the World-Wide Web and other social networks involving millions of nodes. [See e.g. Easley and Kleinberg (2010) for a college level introduction to applications of network theory. For a more advanced treatment see Newman et al. (2003).] While our goal is to focus on the history and not the mathematics of innovation networks, there are some basic ideas and properties of graphs that will be useful for our discussion.

Below is an example of an abstract graph (Fig. 1.8) that can be represented either as a pictorial object with nodes and links or as a matrix (Fig. 1.9). The rows and columns of the matrix are labeled with the names of the nodes (circles). The influence matrix (Fig. 1.9 Top) is constructed as follows: If there is a directed link from e.g. node D to node C, we place a '1' in the D row in the 'C' column. If there is a bidirectional link between A and B, we place a '1' in the A row in the B column and place a '1' in the B row in the A column. The diagonal of the matrix has zeroes and if there are no links between nodes the corresponding entries are zeroes. To construct the Adjacency Matrix in Fig. 1.9 bottom, we assume that all the links are bidirectional and the matrix becomes symmetric.

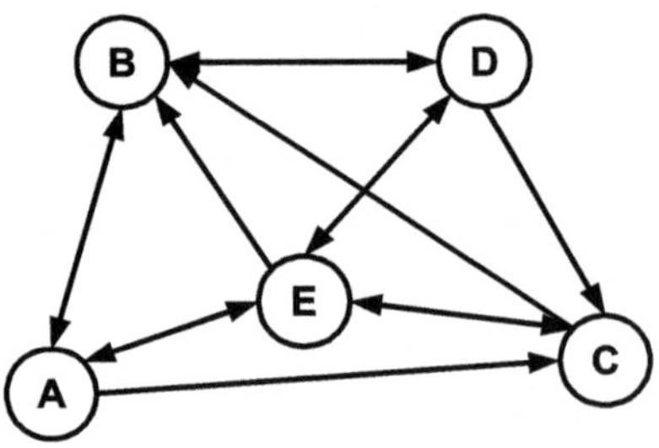

Fig. 1.8 Five-node network with one-way and two-way links

Fig. 1.9 **a** *Top*: Influence matrix for the example in Fig. 1.8. **b** *Bottom*: Adjacency matrix for network example in Fig. 1.8a. Note the symmetry

(a)

Nodes	A	B	C	D	E
A	0	1	1	0	1
B	1	0	0	1	0
C	0	1	0	0	1
D	0	1	1	0	1
E	1	1	1	1	0

(b)

Nodes	A	B	C	D	E
A	0	1	1	0	1
B	1	0	1	1	1
C	1	1	0	1	1
D	0	1	1	0	1
E	1	1	1	1	0

The influence and adjacency matrices for the graph in Fig. 1.8 are shown in Fig. 1.9.

One of the measures we seek to discover in HIN is the average distance between different nodes. For example in the graph of Fig. 1.8, the 'distance' between node A and D is two links. In modern social networks, we call this the *degrees of separation* (see e.g. Watts 2003). For the steam engine network represented by the matrix in Fig. 1.7 we find a distribution of separation paths in this network shown in Fig. 1.10. For example the 'distance' between Watt and Benjamin Franklin in Fig. 1.5 is two links while the distance between Watt and Oliver Evans is four links. Most of the nodes in Watt's network are only separated by 1,2 or 3 links. Algorithms in graph theory that address this problem are called 'shortest path' algorithms and use the adjacency matrix similar to that in Fig. 1.7 (see e.g. Wilson and Watkins 1990 , Sect. 8.2).

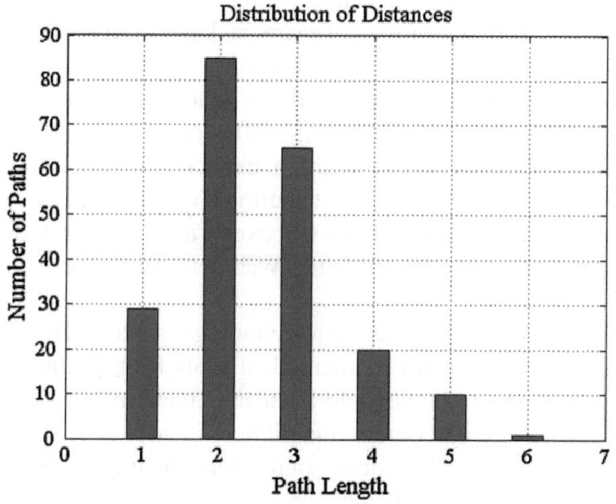

Fig. 1.10 Distribution of number of links in the path between any two nodes for the steam engine network surrounding James Watt in Fig. 1.5 and matrix in Fig. 1.7

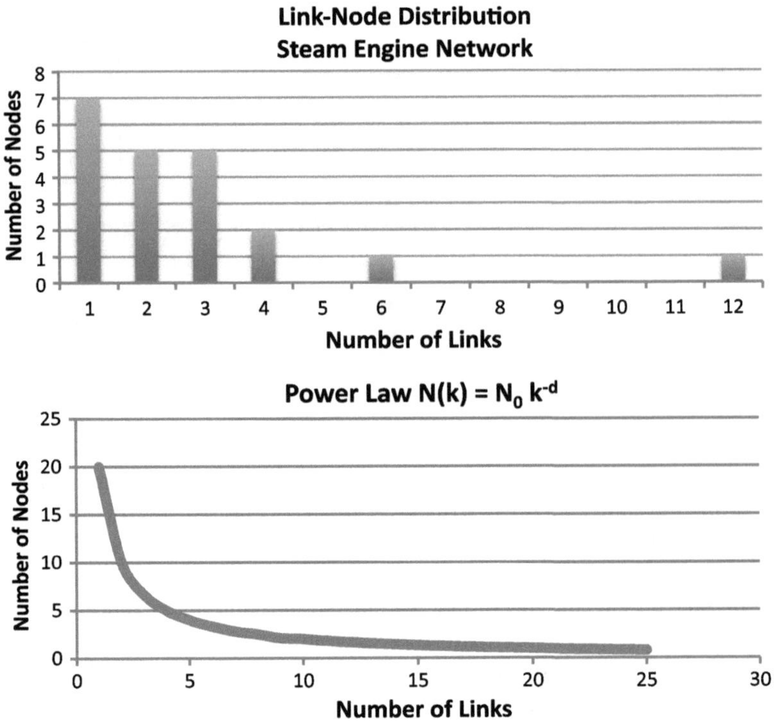

Fig. 1.11 *Top*: Distribution of the number of links per node for the steam engine network of Fig. 1.5. *Bottom*: Power law function for d = 1

The data in Fig. 1.7 was generated with a computer algorithm *'graphshortestpath'* in the Bioinformatics Toolbox of the software package MATLAB.

A classic property of networks is the distribution of number of links. In the steam engine network Fig. 1.5, there are a few nodes with a relatively high number of links and many nodes with few or only one link, as illustrated in Fig. 1.11. If we increase the number of nodes in this network, we might expect to fit a continuous curve drawn through the points in the graph. Such a distribution is called a *scale-free network* since there is no characteristic number of links between nodes (see e.g. Barabasi 2003). Modern networks such as the World Wide Web also have a scale free property of links and nodes.

Thus combining historical studies of a technology, such as Thurston's history of steam power, along with modern mathematical tools from graph theory and computer algorithms [e.g. the influence and adjacency matrices], the modern historian of science and technology can obtain quantitative measures of the creation of new technical and scientific ideas. This does not displace the traditional narrative history methodology, but enhances and extends the understanding of the societal role in the creative process in science and technology.

Compared to the historical networks in this book, with only a few dozen nodes and perhaps a few hundred links, the study of modern social networks such as the World-Wide Web has millions of nodes and links. Modern network analysis affords an entire suite of statistical measures to characterize these large networks as described in a review paper by M.E.I. Newman (2003). These mathematical tools measure network properties such as clustering, path length distributions between nodes as in Fig. 1.10, mixing patterns, so-called 'small world' effects, degree correlations and more. We shall not pursue these advanced properties of networks here. First this is an introductory monograph on history of invention, and second the small number of nodes and links in our networks do not make for statistical rigor. Nonetheless, we will seek to use some of the simplest tools of general network theory as illustrated above. Readers seeking deeper knowledge of networks should consult the works of Newman, Barabasi, Watts and their colleagues. (See Newman et al. 2003 for a collection of research papers surveying modern network theory, or the introductory college textbook of Easley and Kleinberg 2010).

1.6.1 Growth of a Network Knowledge Commons: Timelines and Social Networks

One of the classic tools of historians is the *event timeline graph*. Important events in an historical subject are marked on a horizontal or vertical time axis similar to the chart in Fig. 1.12. In proposing a network model for innovation, we assume that technical and scientific information accumulates in a subject area, such as steam engine technology, and subsequently influences nodes and events in the future. Therefore we show a sample timeline chart with a simultaneous growth of an associated social network. Since accumulated knowledge has value for the network, in Fig. 1.13 we show the integrated number of events with time for an exponential growth.

We propose in this book that the exponential growth for different innovation fields, are correlated with the growth in the network. Accumulated knowledge feeds the growth of the network and vise versa. Thus one of the tools for studying growing networks, is the integrated measures of innovation information. In modern scientific fields, we can measure the growth in published research papers. For earlier times we must seek other measures such as the number of patents filed or number of machines built over a given time period.

One measure of the accumulated network knowledge is to look at the density of events in the history of a scientific or technical field. In the case of the steam engine, new materials were used, new topologies and mechanical components were developed and new thermodynamic concepts were employed as the steam engine evolved. Also new engines were built and tested in the field and this experience gave confidence to future contributors to the field. Each new machine also created a competitive incentive for future designers and entrepreneurs to build a bigger and better machine. This accumulation of knowledge and experience contributed to the

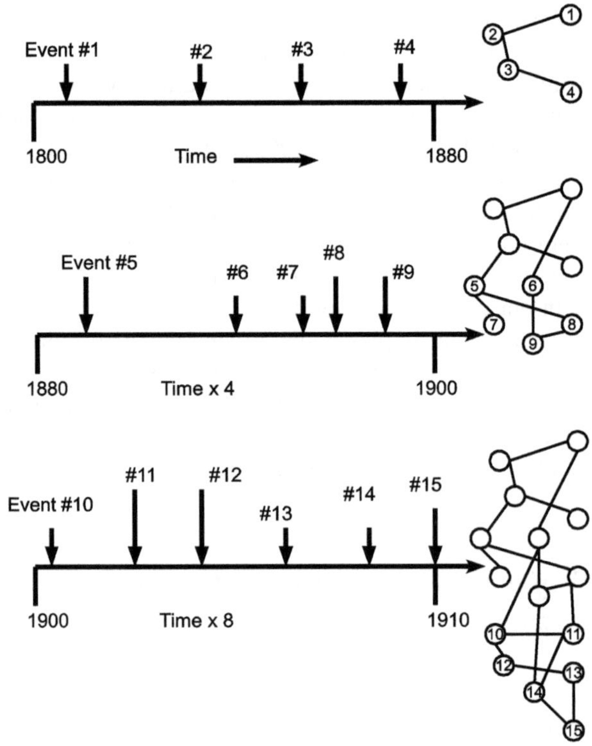

Fig. 1.12 Sample event timeline and growth of associated innovation network. The growth of the number of events and/or nodes versus time is plotted in Fig. 1.13a, b

dynamics of the growth of the network. One could plot the frequency with which new technical events were happening. However to smooth out the noise and randomness in the event history, we propose to integrate or add the number of events in a function called *G(k)* from time t_0 to time t_k. In mathematical notation we would write;

$$G(k) = \sum ß_i e_i; \{i = 0, 1, 2, 3 \ldots k\}$$

Here $\sum$ denotes a summation operation, the event numbers $\{e_i\}$ are all equal to '1', and the event weights $\{ß_i\}$ range from zero to one, i.e. $0 < ß_I < 1$. If an event is very important then $ß_I = 1$, if an event has marginal impact on the field, then $ß_I = 0.1$. If a new technology is advancing very rapidly, then when *G* is plotted in time it will exhibit a sharp rise, which we propose is also indicative of a rapid growth in the innovation network. A hypothetic example is shown below in Figs. 1.12 and 1.13. In Fig. 1.13b, the vertical scale is stretched into a logarithmic scale and the curved exponential curve in Fig. 1.13a looks closer to a straight line. This is a technique used by experimenters to quickly see if data appears to be an exponential function of time, rising very rapidly in time.

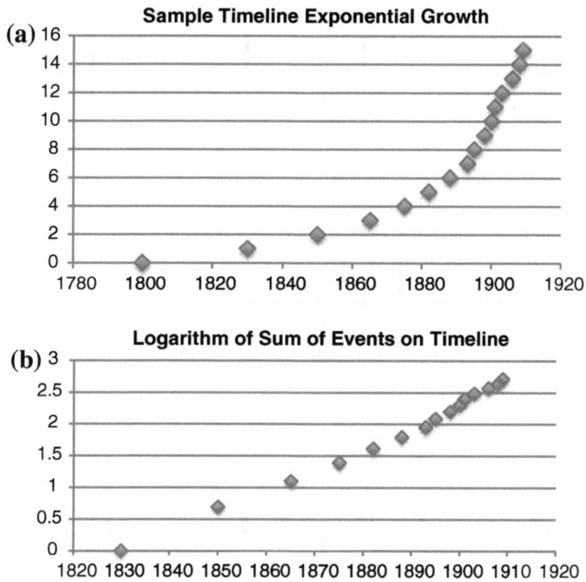

Fig. 1.13 **a** Graph of number of nodes or timeline events for a sample innovation field versus time for exponential growth. **b** Plot of Logarithm of the number of events versus time. The bi-linear structure of this data suggests exponential growth with lower rate in the early years and faster exponential growth of the field in later years

Historical data for the growth of steam engine technology is shown in the figures below using timeline data from Thurston (1878). Figure 1.14 illustrates the partial event timeline for the steam engine and the accumulated number of events as a measure of the steam engine knowledge commons is plotted in Fig. 1.15. Another measure of innovation growth in the use of steam power is the number of units in use in the UK, shown in Fig. 1.16.

Can one construct a model for the growth of innovation networks? Is it possible to predict when a field or product will undergo exponential growth based on early observation of the network dynamics? Using a timeline of events in the history of the steam engine from Thurston (1878) we see the increase in the number of events in the development of the steam engine in Fig. 1.15. We assert that this is proportional to the increase in the nodes in the network diagram of Fig. 1.5 similar to that in Fig. 1.12. In the growth of the integrated events curve of Fig. 1.15a, we can see an initial linear growth in the early part of the eighteenth century and then a more rapid rise in the later half of the eighteenth century. A similar sharp rise in the number of steam engines produced in the UK is shown in Fig. 1.16. It is postulated that the increase of innovation node growth and steam engine product growth are coupled. That is *behind every new emergence of a technical product is a growing social network associated with that technology.*

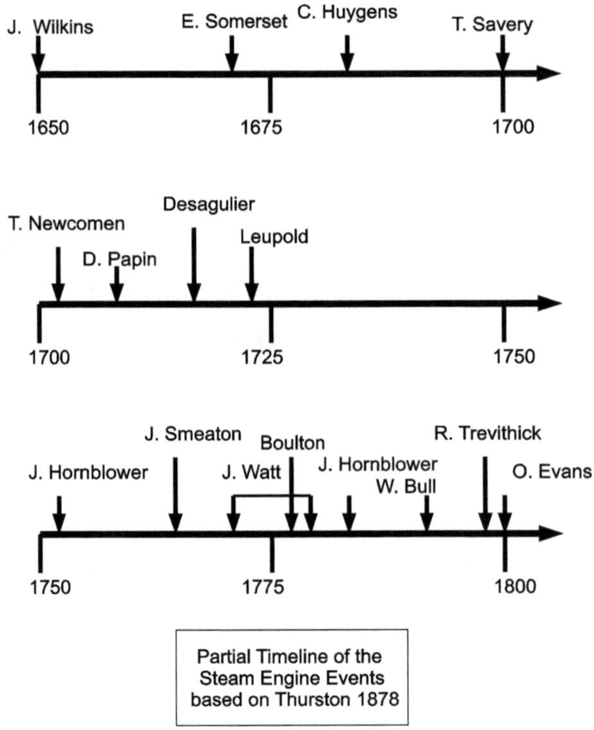

Fig. 1.14 Partial timeline for significant events in the development of the steam engine based on Thurston's history book of 1878

As we shall see in other examples, this behavior seems to be a typical pattern in the evolution of technical and scientific ideas. Initially there is a steady or linear growth of players or nodes that at some time begins to rise in a nonlinear or exponential manner. It is the recognition of the initial time of this exponential rise that makes millionaires out of investors of new technologies.

We must use caution in applying such analysis to new technologies in that the rise in the product penetration into the market, as shown in Fig. 1.16, usually lags the rise in the developmental events as illustrated in Fig. 1.15a for the steam engine. It is usually easier to measure production, but it is more difficult to discover the moment of exponential growth of the social network that is responsible for the new product or service.

In biology one is familiar with the maxim that without mortality factors, populations grow in proportion to the population. Translated into mathematical terms such growth is called *exponential* increase. The formulation of mathematical models of Malthusian population growth can be dated to Verhulst in 1846 (Usher 1929). Similar growth functions are found in chemical and nuclear physics. Such models embody the idea that with more elements or nodes in the system, more will interact and create more nodes.

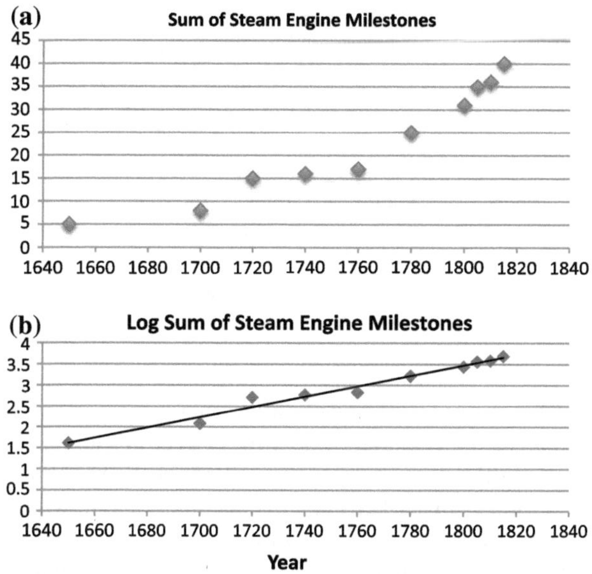

Fig. 1.15 **a** Growth of number of events in the steam engine network 1700–1800 of Fig. 1.14 based on Thurston (1878). **b** Logarithm of **a**: linear line suggests a single the exponential function will describe the growth rate

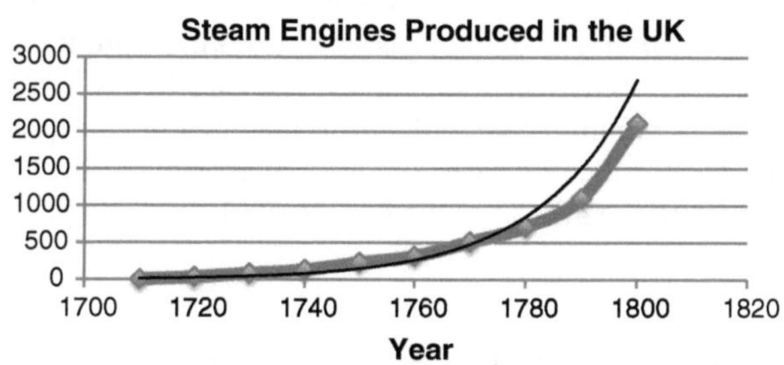

Fig. 1.16 Exponential growth in the number of steam engines produced in the UK from the time of Newcomen [1712] to the end of the Watt–Boulton patents, 1800. The data includes machines from over a dozen manufacturers (Kenefsky and Robey 1980). Solid line is the best fit exponential function using EXCEL

Of course populations cannot grow to infinite size due to limitations of food, energy and resources as well as predatory behavior of other species. In bio-mathematics there is the famous two-specie model of the 'predator and prey' in which as a primary specie grows, it provides more food for the predator species. As the predator consumes the primary species and depletes its own food supply, the predator population declines allowing the primary group to increase its numbers again. Both species exhibit oscillatory population histories but out of phase with each other.

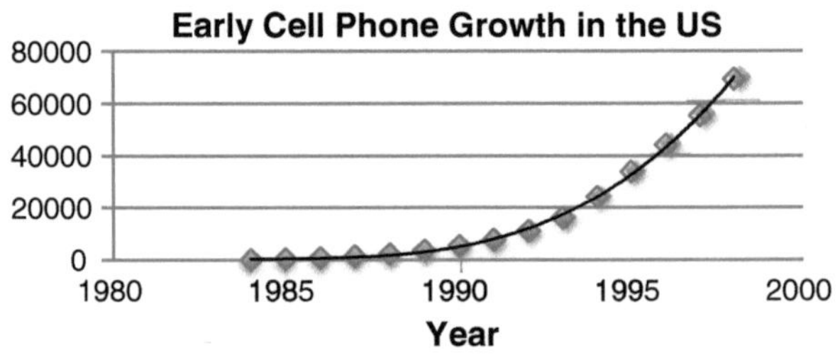

Fig. 1.17 US Cell Phone Subscribers from 1985–1998 (Data from *The World Almanac, 2000,* page 628) Exponential curve fit using EXCEL

The history of technology and invention has many cases of exponential growth and decline, similar to the population growth models in biology (see e.g. Rogers 1962). For example in the early days of the cell phone, 1985–1998, the number of customers who became a subscriber grew exponentially as shown in Fig. 1.17.

When social networks are attendant to the fast growth of a new technical idea or scientific theory, the rate of node increase is often proportional to the number of nodes, and the number of links grows in a similar manner.

Earlier generations of historians of technology such as Usher (1929) have asked similar questions; 'How do new things happen?" In his seminal *History of Mechanical Inventions,* the opening chapters addresses the so-called 'great-man' theory of invention *theory of invention*. Usher discusses humankind's propensity for "novelty in thought and action". Thus when human societies are free to embrace new ideas their natural desire to communicate will spread that idea in an exponential manner. The thesis in this book is that the early exponential growth of new technologies and innovations is linked to the growth of a social network attendant to that innovation.

1.7 Mathematical Models for Network Growth

Given the sharply rising curves associated with the advance of the steam engine in Figs. 1.15 and 1.16, it is tempting to postulate a mathematical model for the increase in nodes in a historical innovation network. In general we would imagine a sequence of node growth, $\mathbf{N}_\mathrm{n}$, along with a set of links, $\mathbf{L}_\mathrm{n}$, where 'n' is a measure of the passage of time, and the next node set $\mathbf{N}_{\mathrm{n}+1}$, $\mathbf{L}_{\mathrm{n}+1}$, are functions of the previous number of nodes and links. This general relationship might be written in the following equations:

Node Growth

$$\mathbf{N}_{\mathrm{n}+1} = \mathbf{N}_\mathrm{n} + \mathbf{G}\{\mathbf{N}_\mathrm{n}, \mathbf{L}_\mathrm{n}\};$$

Link Growth

$$\mathbf{L}_{\mathrm{n}+1} = \mathbf{L}_\mathrm{n} + \mathbf{F}\{\mathbf{N}_\mathrm{n}, \mathbf{L}_\mathrm{n}\}$$

The functions **G** and **F** above can be either deterministic or based on probability theory. In modern theories of social networks such as Facebook, where millions of transactions are occurring every day, probability models are used to predict growth. Attempts to predict the growth of historical innovation networks or modern technology networks might need another approach than probabilistic models since the number of nodes is often orders of magnitude lower than networks such as the World-Wide Web.

Models for the growth of networks in sociology date back to the 1950s and 1960s in especially the work of Derek de Solla Price (1965a, b), mentioned above. These models often assume that the increase in the number of nodes and links is proportional to the current number of nodes and links. Modern growth models have been proposed by Barabasi and Albert (1999). These and other models are reviewed by Newman (2003).

In this book however our focus is on the development of technical and scientific ideas that result from the evolution of a social network. Thus we are interested in the slow rise and rapid growth stages of a science and technology. To capture this phenomena we propose a network node growth that is proportional to a constant plus a term proportional to the number of nodes; i.e. if we define

$$\Delta \mathrm{N} = \mathrm{N}_{\mathrm{n}+1} - \mathrm{N}_{\mathrm{n}}$$
$$\Delta \mathrm{N} = [a + b\mathrm{N}]\Delta t$$

For early times in this process, this model predicts that N(t) growth is proportional to time i.e. N = at.

For later times, this model predicts that N(t) grows exponentially i.e.

$$\mathrm{N} \Rightarrow \mathrm{N}_0 e^{bt}$$

An example of such a node growth model for a = 2, b = 1/4 is shown in Fig. 1.18 where the time unit is decades. The pure linear model with a = 2 is one in which the number of nodes doubles each decade. The combined linear and exponential model shown as the top curve increases significantly faster than the linear growth curve

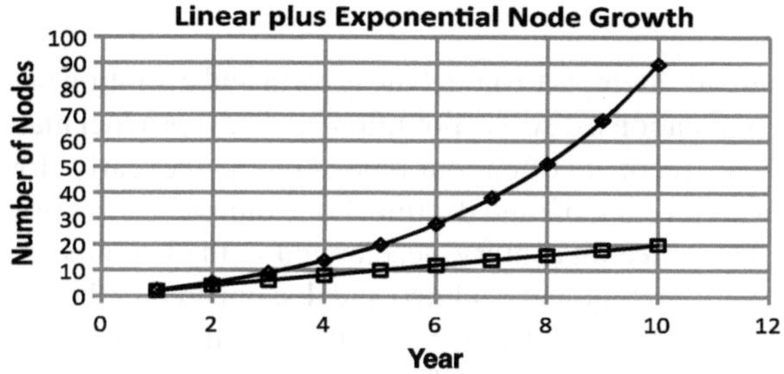

Fig. 1.18 Node growth model: *Bottom*: Linear growth; *Top*: Linear plus exponential growth $a = 2$; $b = 1/4$

below. The increase in the network nodes is proportional to the number of nodes as well as time.

Some technical developments such as ballooning in the nineteenth century [Chap. 4] exhibit mainly linear growth in time, while others such as radio development exhibit exponential phenomenon [Chap. 5]. Linear growth implies that the historical network does not play a strong role. New players come into the field because of their inherent interest and not because others are in the field. On the other hand, when new players are inspired by the aggregate of players or nodes in the network, the network grows more rapidly and appears exponential. Exponential growth is like a chain reaction where the larger the number of nodes the larger the growth rate in the network. We will find this phenomenon in many of the innovation examples in this book.

As we shall see in Chap. 4 on aviation, as historical networks advance, not only does the number of nodes grow, but so also does the number of links between the nodes. And in the age of the telegraph and the early telephone age, this growth in both the nodes and links occurs in an exponential way.

Not all examples of historical innovation undergo exponential growth. In the next chapter we show that in the period between the Renaissance and the Industrial revolution, there is evidence of growth in machine technology knowledge but it grew in a linear manner, most likely because the information and transportation networks that underpin modern technological development were not in place until the nineteenth century.

The study of the dynamics of technological change has been an active field for over a half century not only in history of technology and sociology but also in economics (see e.g. Mansfield 1968). It is beyond the scope of this book to survey all of these sources. Econometric models of technological dynamics often begin with statistical data. Our approach in this book is history based, beginning with the minutia of historical biographies and surveys of technology to find nodes and links and then applying statistical and mathematical methods where possible.

1.7.1 Growth Models and the "S" Curve

There have been many attempts to model the growth and evolution of observable phenomena such as populations, disease, pollution, and new product market penetration. In these phenomena, there are often four phases of change; early slow growth, rapid exponential growth, leveling off and decline. For example the use of telegrams began in the early to mid nineteenth century, accelerated in the late nineteenth, leveled off in the early twentieth century and declined and disappeared in the late twentieth century. One mathematical model that captures the first three phases is the "S" curve, shown in Figs. 1.19 and 1.20 (see e.g. Rogers 1962; Bejan and Lorente 2011). This model is often useful in capturing product or service penetration in a market for a new drug or a new fashion craze. A mathematical curve that has a similar shape is the sigmoid function plotted in Fig. 1.19.

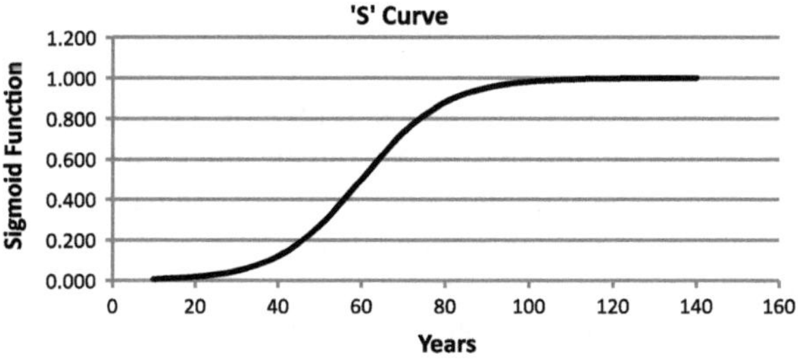

Fig. 1.19 Graph of 'S' Curve or sigmoid function $= [1/(1+exp\{-t+60\})]$; this function shows slow growth, then rapid acceleration and saturation

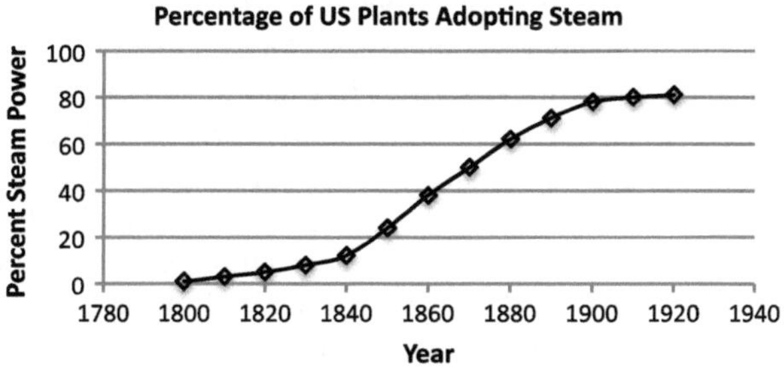

Fig. 1.20 An 'S' curve. Percentage of manufacturing plants in the US that replaced water power with steam power, 1800–1920. (From Atack et al. 1978.)

In technology, new product numbers first rise, then level off as the ultimate number of potential customers is finite and as new products ('predators') replace the customer need for the original technology (Rogers 1962). This phenomenon is sometimes called an "S" curve. An example of S-curve growth for the use of steam power in the United States is shown in Fig. 1.20. The authors noted that conversion to steam was greatest in the Midwest where waterpower was scarce and made least progress in New England where there was a long history of hydropower-based manufacturing. Hence the asymptote in Fig. 1.20 is 80% steam power.

The 'S' curve has mathematical roots in the modeling of population growth. Pierre-Francois Verhulst [1804–1849] used such a model, in 1838, to describe the constraints on human populations. In such a model the increase in the population or number of nodes in network theory ΔN is assumed to be proportional to both a linear and quadratic function of N;

$$\Delta N = aN(1 - N/N_0) \quad \Delta t.$$

The nonlinear term N^2, represents the constraints on the expansion of the population when it grows too large. When ΔN and Δt are small, the limit of this expression is a differential equation known as the *logistic equation*. Its solution where N is given as a function of time takes the form of the sigmoid function in Fig. 1.19.

$$N = N_0/(1 + e^{-at}).$$

The Nobel Prize winner Ilya Prigogine has written of self-organizational principles in natural and societal systems and uses the logistic equation to model the growth of insect colonies as well as the economic growth of cities (Nicolis and Prigogine 1989) Economists have also used the logistic equation to describe market penetration of a new innovation (see e.g. Silverberg et al. 1988).

Literature in the history of technology contains many examples of technological growth (see e.g. de Solla Price 1965a, b, 1951). In this book we present evidence that such increases arise from the emergence of social networks between inventors, artisans, scientists, educational institutions and industrial organizations.

As technologies mature, the role of the person-node diminishes and the importance of the manufacturing company node increases. An example of this is the growth of the internal combustion industry described in Chap. 3. Although personalities such as Otto and Daimler were important nodes in the early period say 1860–1890, by the turn of the century there were many licensed companies producing engines. This transition was enabled by a machine tool infrastructure already in place as a result of the earlier steam engine technology and its role in building railroads, steamships and manufacturing technology. A similar scenario existed in the evolution of wireless and radio technology, described in Chap. 5, where the existence of both telegraph and electric generation technologies allowed mature companies to move into the new radio age with relative ease.

The transition of HIN based on individual nodes to networks based on institutional nodes such as research universities and corporations would be an interesting study but is beyond this book except anecdotal examples.

1.8 Lessons from History Innovation Networks

Anticipating the results of succeeding chapters, the case studies of innovation social networks described in this book suggest some lessons for modern innovation.

(i) New ideas in science and technology are often the product of both evolution and complex social networks.
(ii) New technical artifacts are usually preceded by the emergence of other technologies, e.g. the internal combustion engine preceded aviation.
(iii) Secrecy and patents do not seem to inhibit the exponential growth of new ideas.
(iv) Both physical communication and transportation infrastructure networks are necessary to the growth of innovation social networks.
(v) Although artisans played a major role in late eighteenth and early nineteenth C. technologies, educational and research institutions have become important nodes in modern innovation networks.
(vi) Behind every emergence of a new technical product is a growing social network associated with that technology.

The succeeding chapters expand on these themes through detailed discussion of specific examples of innovation and invention.

In contrast to the romantic view of scientific creativity, modern engineering education stresses the collaborative nature of technical design. The engineering team approach as a tool for creating new robotic machines is now entrenched in many high schools and engineering colleges. Student team competitions in robotics, mathematics, efficient home design and racecars are all the rage in engineering education today. The irony of course is that many of the young people attracted to these teams are also inspired by the genius stories of invention. Many see themselves as becoming the hero or heroine inventors of the future.

One bastion of the isolated scientist model is the PhD research dissertation system. But in many areas of science, such as gene theory and high-energy physics, there are teams of up to hundreds of biologists and physicists involved. Recent examples are the Large Hadron Collider Project at CERN in Switzerland or the Mars planetary missions of NASA. [CERN boasts that 10,000 scientists or half the world's particle physicists come to CERN to do research.] Many admit that modern science has become a global collaborative effort but still harbor the dream of the Nobel Prize; the ultimate embodiment of the 'genius' model of scientific innovation. Berkun (2010) in *The Myth of Innovation* makes a case for societal and organizational roles in the creation of new technical ideas and products.

1.8.1 Self-Organized Critical State Models for Innovation

In Fig. 1.11 we presented steam engine statistics for the number of nodes N(k) with k links as a power law function $N = N_0 k^{-d}$. This distribution is similar for the World-Wide Web. Statistics for earthquakes, forest fires, the growth of cities, all have power law properties similar to modern networks such as the World-Wide Web. For earthquakes, N is the number of quakes over time with strength 'k'. For demographics, N is the population and 'k' is the rank from high to low. In the last two decades physicists have attempted to link these phenomena to similar behavior in the physical world of science such as the properties of sand piles, avalanches and magnetic domains in materials. (Buchanan 2000) Physicists have tried to apply a mathematical theory called *self-organized criticality* or SOC, to such phenomena (Bak 1996). The appellation '*criticality*' refers to the possibility that the system will experience a significant event such as an earthquake or an avalanche in a sand pile. '*Self-organized*' implies that after such an event, the system resets itself to eventually trigger another event. The system is said to be on the edge between order (no significant events) and chaos (unpredictability of events)

Recently economists have tried to find evidence in the dynamics of financial markets to suggest SOC response in the social behavior of markets. In this book we present data for the evolution of technical innovation that exhibits sudden exponential growth in measures of technical progress such as patents filed and market use of the technology. The construction of social network diagrams has associated with it similar statistics regarding the growth of network nodes and links, namely the sudden exponential increase in the size of the network. It is tempting to suggest that technical innovation is a case of *self-organized criticality* in which the social network of contributors advance the technology.

One broad outline of such a theory imagines the social innovation network as a set of players in a technical innovation market. When some new technology comes along, such as the steam engine or internal combustion engine, players on the sideline with relevant knowledge or capital begin to come into the market by connecting with other players. We might assume that there is a reservoir of technical knowledge, generated from other innovations, that can at some point be transferred to the new technology. As the network grows through linkages, new players come into the network with new knowledge or capital that in turn stimulates more linkages with other players in the knowledge reservoir.

For example, by the time the Wright brothers came into the aviation network c. 1900, there were already several dozen significant nodes, including Lilienthal, Langley and Chanute. The internal combustion network was already in a critical state with hundreds of players, dozens of manufacturing firms and several thousand players associated with ballooning technology. We could argue that the Wright brothers were drawn into the early aviation network. And after the Wright brothers created their own flying machine of 1903, they in turn drew many more players into the aviation social network. [See Chap. 4 for details.]

In another example discussed in Chap. 3, by the time Henry Ford produced his first horseless carriage c. 1896, there were more than 500 patent applications in the US alone related to automotive technology. And when Ford produced his Model T , c. 1903, he induced hundreds more players to join the automotive innovation network.

The goal of this book is not to create a SOC theory for social innovation networks but to provide historical data in a statistical format that might suggest models for innovation based on the growth of historical social networks.

References

Atack J, Bateman F, Weiss T (1978) The diffusion and adaptation of the steam engine in manufacturing industry during the 19th century. Indiana University Faculty Working Paper No. 515, Sept. 27

Bak P (1996) How nature works. Copernicus Press, Ithaca

Barabasi A-L (2003) Linked. Plume-Penguin, Cambridge

Barabasi A-L, Albert R (1999) Emergence of scaling in random networks. Science 286:509–512

Basalla G (1988) The evolution of technology. Cambridge University Press, Cambridge

Bejan A, Lorente S (2011) The S curve. J Appl Phys 110:024901. (See, Mechanical Engineering 2012)

Berkun S (2010) The myth of innovation. O'Reilly Media, Sebastopol

Buchanan M (2000) Ubiquity. The science of history. Weidenfeld and Nicolson, London

Carnegie A (1905) James Watt. Doubleday Page & Co., New York

Collins R (1998) The sociology of philosophies: a global theory of intellectual change. Harvard University Press, Cambridge
Dickerman L (2013) Inventing abstraction 1910–1925. How a radical idea changed modern art. Museum of Modern Art, New York
Easley D, Kleinberg J (2010) Networks, crowds and markets: reasoning about a highly connected world. Cambridge Univ, Press, Ithaca
Farey J (1827) A treatise on the steam engine. Longman, London
Gladwell M (2000) The tipping point. Little Brown and Co, New York
Hughes TP (1983) Networks of power. Electrification in Western society. Johns Hopkins Press, Baltimore
Keats-Rohan KSB (ed) (2007) Prosopography, approaches and applications: a handbook. Univerisity of Oxford, Oxford
Kenefsky J, Robey J (1980) Steam engines in 18th century Britain: a quantitative assessment. Technol Cult 21(2):161–186
Koetsier T, Ceccarelli M (2012) Explorations in the history of machines and mechanisms, chap. 1. Moon FC. A social network model of early aviation history. Springer, Dordrecht
Kragh H (1987) Introduction to the Historiography of Science. Cambridge University Press, New York
Lardner D (1836) The steam engine familiarity explained and illustrated. E.L. Carey and A. Hart, Philadelphia
Mansfield E (1968) The economics of technological change. W.W. Norton, New York
Mokyr J (1990) The lever of riches: technological creativity and economic progress. Oxford University Press, New York
Moon FC (1992) Chaotic and fractal dynamics. Wiley, New York
Moon FC (2007) The machines of Leonardo da Vinci and Franz Reuleaux. Springer, New York
Moon FC (2012) A social network model for innovation in early aviation history (see Koetsier M. and Ceccarelli T)
Newman MEJ (2003) The structure and function of complex networks. SIAM Rev 45:167–125
Newman MEJ, Barabasi A-L, Watts DJ)(ed) (2003) The structure and dynamics of networks. Princeton University Press, Princeton
Nicolis G, Prigogine I (1989) Exploring complexity (See Chap 6. W.H. Freeman, NY)
North Dakota State University (2000) Mathematics Genealogy Project, Genealogy Mathematical NDSU Education. Department of Mathematics and American Mathematical Society
Preiser-Kapeller J (2010) Calculating Byzantium?. Social network analysis and complexity sciences as tools for the exploration of medieval social dynamics. Austrian Academy of Sciences Report
Rogers EM (1962, 1995) Diffusion of Innovations, 4th edn. The Free Press, Simon & Schuster, New York
Rosen W (2010) The most powerful idea in the world. Random House, New York
Rosenberg D, Grafton A (2010) Cartographies of time: a history of the timeline. Princeton Architectural Press, New York
Schmidt-Burkhardt A (2011) Maciunas' learning machines: from art history to a chronology of fluxus, 2nd edn. Springer, Wien
Silverberg G, Dosi G (1988) Innovation, diversity and diffusion: a self organization model. Econ J 98(393):1032–1054
de Solla Price D (1963) Little science big science. Columbia University Press, New York
de Solla Price D (1965) Is technology historically independent of science: a study in statistical historiography. Technology and Culture, Autumn 6(4):553–568
de Solla Price D (1965b) Networks of scientific papers. Science 149(3683):510–515
de Solla Price D (1951) Qualitative measures of the development of science. Arch Int 14:85–93
Tann J (1981) The selected papers of Boulton and Watt, vol I. MIT Press, Cambridge
Thurston RH (1878) A history of the growth of the steam engine. D. Appleton and Co., New York

Uglow J (2002) The Lunar Men. Faber and Faber, London
Usher AP (1929) A history of mechanical inventions. Dover, New York (1982)
Watts DJ, Strogatz SH (1998) Collective dynamics of small world networks. Nature 393:440–442
Watts DJ (2003) Six degrees. Norton, New York
Wilson RJ, Watkins JJ (1990) Graphs: in introductory approach. Wiley, New York

Chapter 2
Networks in the Machine Age: From Leonardo to Clocks to Reuleaux

Abstract We apply a network analysis to the history of kinematics theory of machine design from the fifteenth to the nineteenth centuries beginning with Leonardo da Vinci. As a particular example we construct a network chart for the evolution of precision marine clocks. We also look at the social network of Franz Reuleaux and other kinematics contributors of the nineteenth century. The kinematics network has links to the internal combustion network of Chap. 3.

The modern origins of the Machine Age of the nineteenth century began in the Renaissance in fifteenth century Italy. In Siena and Florence, artist-engineers such as Mariano Taccola (Scaglia 1984), Francesco di Georgio Martini (Weller 1943), and Leonardo da Vinci (Galluzzi 1997) produced collections of drawings of hundreds of machines and machine elements. An older contemporary of Leonardo, Francesco di Giorgio published his machine drawings, a copy of which was in the library of Leonardo. (Francesco di Giorgio Martini 1450) The manuscripts of Leonardo di Vinci were broken up and were subsequently dispersed throughout Europe. Some of the more famous Leonardo manuscripts that contain machines are the *Codex Atlanticus* in Milan, *Manuscript B* in Paris, and the *Codex MadridI* in Madrid. The *Codex Madrid* is unique in that it marks the first attempt to deconstruct machines into basic machine elements or mechanisms, or what Leonardo called "elementi macchinali".

Recognition of Leonardo's work in science and technology emerged slowly at first though reproductions of his drawings in the early nineteenth century and then accelerated at the end of the nineteenth century with publication of facsimiles of the *Codex Atlanticus* in Milan and the Leonardo Notebooks in the Institute of France. As reproductions of Leonardo's work became commonplace, his reputation as a genius inventor and scientist also spread so that the helicopter, airplane, robots, war machines, clock escapements and a hundred other machine creations are attributed to his unique intelligence by the popular press. But was Leonardo singular in the drawings and designs he put in his notebooks or was he a skillful draftsperson who copied what he saw around him? There are partisans on both sides of the question.

F. C. Moon, *Social Networks in the History of Innovation and Invention*,
History of Mechanism and Machine Science 22, DOI: 10.1007/978-94-007-7528-2_2,
© Springer Science+Business Media Dordrecht 2014

Fig. 2.1 Portrait of Leonardo da Vinci [1452–1519]

In this chapter we will attempt to describe Leonardo's network in fifteenth century Florence and Milan (Elements of this chapter are taken from Moon (2007)).

Shortly after the *Codex Madrid* was rediscovered in 1965, the da Vinci scholar Ladislao Reti (1980a, b) translated the text into English and published several popular books on 'Leonardo the inventor'. He showed that da Vinci had attempted to compile a basic compendium of machine elements. To compare Leonardo's drawings of machine mechanisms with modern machine design books, Reti choose the list of machine elements proposed by Franz Reuleaux in his popular, nineteenth century book on machine design, *The Constructor,* first published in 1864 and translated into four languages and four editions. One of our principal themes is that Leonardo da Vinci was a key node in a network of artist-architect-engineers, which passed on an unbroken chain of knowledge on the nature of machines through four centuries (Fig. 2.1, 2.2).

2.1 A Theory of Machines: Kinematics of Mechanisms

The evolution of machine theory has some similarity with biology. By the middle of the nineteenth century, machines had been invented that had their own sources of energy, were mobile, and were undergoing a process of multiplication and evolution, becoming more complex with every new generation. To bring order to the welter of new machines, engineers began earlier to search for ways to codify and classify these devices. Beginning in the sixteenth century, compendia of machines, such as those of Besson (1571) and Ramelli (1588), were organized according to application, such as pumps, manufacturing machines, construction machines etc. This methodology

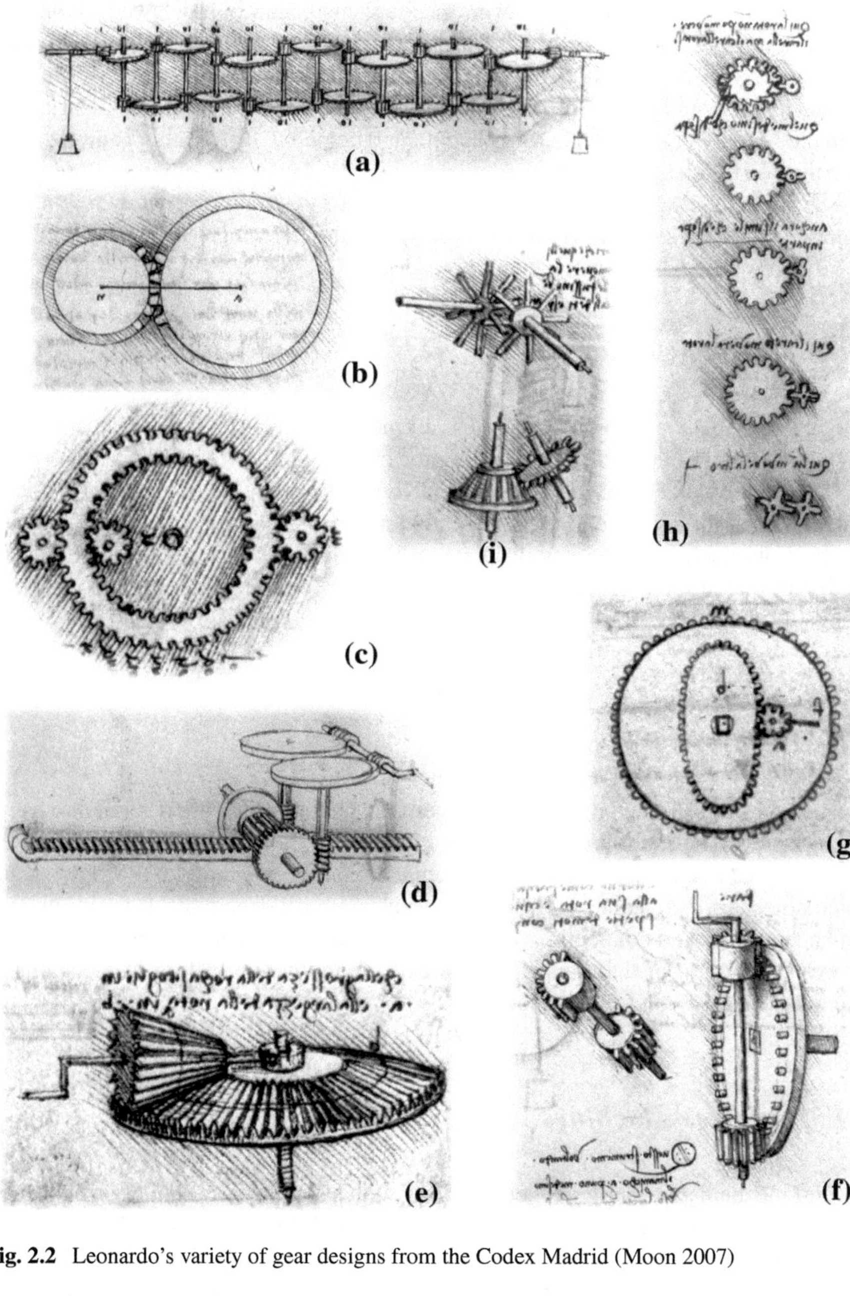

Fig. 2.2 Leonardo's variety of gear designs from the Codex Madrid (Moon 2007)

paralleled similar attempts to organize the biological world such as the *Systema Naturae* of Carolus Linnaeus in 1735.

By the nineteenth century however, mechanisms began to be viewed as *motion-changing devices* and this path was formally codified by Charles Babbage of computer fame, Robert Willis of Cambridge and later by Franz Reuleaux of Berlin.

The creation and design of machines on a rational basis freed the industrial age from the secrecy of the guild and workshop, and helped to diffuse this knowledge all over the world, changing once feudal societies like Japan into emerging industrial powers by the end of the nineteenth century. The following terminology might help in understanding the nature of machine design in the late nineteenth century.

2.1.1 Kinematic Chain

A kinematic chain is a connected series of kinematic pairs (joints: rotary, sliding or prismatic, screw) that form a closed loop or circuit, similar to an electrical circuit. The closed loop of a crankshaft, piston rod, cylinder and cylinder block form one of the most ubiquitous kinematic chains called a *slider-crank chain* found in internal combustion machines. The key property of the kinematic chain is the fact that the motion of one or more of the links in the chain determines the motion of the parts in the rest of the chain.

2.1.2 Mechanisms

Mechanisms are simple or compound kinematic chains which are designed to transform motion. The motion of one link or element in the mechanism determines the motion in the rest of the links in the kinematic chain. For example, the slider crank kinematic chain in a steam engine or an internal combustion engine, i.e. the crank, piston and cylinder and connecting rod, changes the translational motion of the pistons into rotary motion of the crankshaft. Clocks are another collection of mechanisms employing gearing and escapements.

2.1.3 Complex Machines

Several mechanisms coupled together, along with a source of motion or energy, form complex machines. An example is the first automobile of Benz [1885]. It consists of several kinematic pairs (i.e. cylinder and cylinder block, or bevel gear pair) and coupled kinematic chains (e.g. the slider-crank formed by the piston or the chain drive coupled to the friction wheels on the ground), along with the chemical reactions of the fuel-air mixture that provide the force to drive the piston.

2.2 Leonardo: Engineer or Artist?

Was Leonardo an artist who dabbled in sketching machines or was he an engineer who painted in his off-hours? There are literally hundreds of books on Leonardo as an artist, though there are less than a dozen paintings attributed to him. In this brief summary we review those aspects of his life related to his work as a royal engineer. There is often debate as to the facts and dates surrounding the life of Leonardo. Many of the dates given here are from the works of Kenneth Clark, Ivor B. Hart (1925, 1961) and Gibbs-Smith (1978). A casual survey of the *Codex Atlanticus* of Leonardo da Vinci will reveal an amazing array of machines and mechanisms. Some of these are listed in Table 2.1 below.

Leonardo was born on April 15, 1452 in the village of Vinci northwest of Florence. His death is recorded as May 22, 1519 in Amboise, France. His was an out-of-wedlock birth whose father Ser Piero da Vinci was a notary who acknowledged Leonardo's patrimony and took him into his home. As a teenager, Leonardo was apprenticed into the workshop of the painter, sculptor and goldsmith, Andrea del Verrocchio [1435–1488] around 1470 where Sandro Botticcelli was also a student. There he learned the skills of drawing, painting, sculpture and perhaps architecture. Leonardo later joined the Guild of St. Luke as a painter at the age of 20 while still living in Verrocchio's studio. In other words, Leonardo was connected to a network of artisans from whom he learned many of his artistic and design skills.

The fifteenth century was also a time of tremendous change and political tension, especially in the Italian states. Two of the most powerful rulers during Leonardo's years in Florence were Cosimo de' Medici [1389–1464] and Lorenzo de' Medici [1449–1492]. Under both rulers art, architecture and literature flourished. Under the reign of Cosimo, the architect-engineer Filippo Brunelleschi completed the great octagonal dome (55 m across) of the Cathedral of Florence in 1436. The movable-type printing press was invented by Gutenberg in 1450, and by 1500 there were nearly 300 printing press workshops in the Italian states (This is important because Leonardo would have had an opportunity to disseminate his ideas and writings had he so chosen). Florence was a center of silk and wool textile manufacturing perhaps inspiring some of Leonardo's designs for textile machines. Italian mariners had taken a new class of sailing ships as far west as the Azores and later in 1492, Columbus would make the most fantastic voyage to the New World. Copernicus was born when Leonardo was 21.

At the age of thirty in 1481, Leonardo wrote a remarkable resume for the Duke of Milan, Ludovico Sforza [1452–1508], also known as 'il Moro'. A draft of this letter in his notebooks, boasts of Leonardo's skills as a military and civil engineer and at the end briefly mentions his considerable skills as a painter and sculptor.

In 1482, Leonardo left for Milan and was appointed "ducal painter and engineer". He spent the next 18 years working for Il Moro, serving as consultant on fortifications at Milan, Pavia, and Vigenano. In 1499–1500, Ludovico Sforza was displaced by the French King Louis XII in his occupation of Milan, and Leonardo left for a brief stay in Venice in 1499 and later returned to Florence in 1500. In Venice he was occupied

Table 2.1 List of Leonardo's machines by application

Manufacturing machines	Source	Construction machines	Source
Lathe	F1059r/381r.b	Cranes	F4r/1v.b
Screw cutting machine	Ms. G 70v	Swing bridge	F855r/312r.a
Rope making machine	F12r/2v.a	Canal dredge	Ms. E 75v
File making machine	F24r/6r.b	Canal lock gates	F90v/33v.a
Drilling machine	F34r/9v.b	Pumps	F20r/5r.b
Lens-grinding machine	F1057v/380r.b	Hydraulic screw pump	F1069v/386v.b
Needle-making machine	F86r/31v.a	Chain-of pots pump	F1069v/386v.b
Thread spinning machine	F1090v/393v.a	Pulley systems	F1102v/396v.g
Textile machines	F106r/38r.a	Windlass	F1112r/400r.a
Book press	F995r/358r.b		
Wine and olive press	F47r/14r.a	*Power and Transmission*	
Mechanical saw	F1059r/381r.b	Boat paddle wheels	F945r/344r.b
		Air turbine wheel	F46a-r/13v.b
Military Machines		Water power systems	F26v/7v.a
Bombs	F33r/9v.a	Spring-propelled cart	F812r/296v.a
Balistica	F145r/51v.b	Power transmissions	F26v/7v.a
Trebuchet	F160a-r/57v.a	Flying machines	F844r/308r.a
Catapults	F150r/54r.a	Human powered Wheel	F1070r/387r.a
Armed chariots	Turin BB 1030		
Breech-loading mechanism	Ms. B 24v	*Measurement Devices*	
Giant Cross-bow	F149b-r/53v.b	Clock escapements	F96r/35r.a
Rapid-repeating crossbows	F182b-r/64v.b	Odometer cart	F1b-r/1r.bk
Steam-gun	Ms. B 33r	Weighing machines	Ms. A 52r
Multi-barrel gun	F157r/56v.a	Drawing compass	F696r/259r, a-b

Sources Codex Atlanticus; Folio: F: New/Old numbers
Paris Manuscripts; Ms. A-M

with some military consulting for the Venitian Senate. He also worked with the mathematician Pacioli who had fled with Leonardo from the troubles in Florence.

Besides undertaking some paintings, a great deal of Leonardo's service to 'Il Moro' was involved in surveying fortifications, devising town plans, drawing plans for canals, organizing pageants and providing advice on military technology. To carry out these tasks there are a number of names in his Notebooks of workers and students in his Milan workshop. These professionals were machine makers, locksmiths and glass craftsmen suggesting that he was surrounded by men who had some experience and knowledge of machines and technical processes.

During his time in Milan, Leonardo traveled to Pavia in 1490 and met the older Francesco di Giorgio Martini [1439–1501] from Siena, a respected artist-engineer who, following the work of another Sienese engineer Mariano Taccola, (Scaglia 1984) had published a widely circulated book with descriptions and drawings of machines for military and construction purposes. From a record of books in Leonardo's library, we know he had a copy of Francesco's book on architecture and machines. Most of Leonardo's own Notebook drawings of machines date from after this time.

Another important contact vis-à-vis his education as a machine designer and engineer, was his friendship with mathematicians such as Fazio Cardano of the University of Pavia and Fra Luca Pacioli who later wrote on arithmetic and geometry. Leonardo drew the illustrations for Pacioli's book *De Divina Proportione* in 1509. The relation between mathematics and machines has its origins in the work of Aristotle, Hero, and Archimedes. The earlier picture books of Taccola, and Francesco di Giorgio however, did not show the precision that was necessary for machine construction until the work of Leonardo whose interest in geometry is clearly visible in his machine drawings.

In 1500, Louis XII of France occupied Milan and Leonardo departed for Venice and Florence and then a tour of duty in Rome with Cesare Borgia, Duke of Romagna. During this time Leonardo produced maps for the construction of canals, another engineering task that involved careful geometric drawing skills. In the 20 years after he had left Florence he had started a number of paintings and sculpture projects, but very few were ever finished or survived. The French who were still in Milan referred to Leonardo as "*nostre peintre et ingenieur ordinaire*" He also spent a short time in Rome from 1514–1516, but was overshadowed artistically by the younger Michelangelo and Raphael.

Although Leonardo made thousands of drawings, sketches, designs and pages of notes on dozens of subjects, we have no record of a published book. There is some historical reference to a book on principles of painting. But in engineering, he did not write or publish any book on the subject for which he later gained universal respect, the design and invention of machines. His drawings, papers and notebooks were bequeathed to a younger pupil, Francesco Melzi [-d.1570] who kept them until his death at which time Melzi's son mishandled them. These papers were later given to the sculptor Pompeo Leoni who tried to reorganize the welter of drawings into codices related to several themes. In this work he often cut out drawings and pasted them into other works. The largest collection of manuscript pages was called the *Codex Atlanticus*, about 4000 folios, which was given to the Biblioteca Ambrosiana, Milan. This large 12-volume work contains hundreds of machine drawings. This work was first published in facsimile in 1894 and can been found in many large libraries around the world.

In different codices Leonardo mentioned the Roman engineer Vitruvius, Archimedes, the Commentaries of Caesar, or the geometry of Euclid. In a remarkable discovery in the recently translated *Codex MadridII*, was a list of "*books I have left locked in a chest*". This list of over 100 books includes works on Aristotle, The Bible in Italian, Letters of Ovid, St Augustine, Justinius, (Roman historian), Fables of Aesop, Petrarch and Pliny. In mathematics, he owned books by Euclid and Luca Pacioli. In architecture and engineering his library contained books by Alberti, Valturio and Francesco di Giorgio as well as one entitled "A book of Engines". There is also one likely to be Philon of Byzantium's "Pneumatics". It is interesting to note that although the Roman engineer Vitruvius is mentioned by Leonardo in several locations in his manuscripts, his work is not listed in the book list in Codex Madrid II. Vitruvius's Book X of his treatise on Architecture was a key reference to machine engineering of Roman and Greek antiquity.

2.3 Influence Network of Leonardo da Vinci in Machine Design

Leonardo da Vinci is sometimes portrayed in contemporary media as a singular genius, especially in the subject of machines and inventions. However as there were artistic influences of earlier painters and sculptors on Leonardo's art, there was also a network of architect-engineers and artist-engineers who were either building or drawing designs for machines and whose work likely had some influence on Leonardo.

To graphically summarize the evolutionary influences of ideas in machine design and invention on Leonardo da Vinci, an influence network chart is shown in Fig. 2.3. At the top representing antiquity, including the work of Archimedes, Ctesibius and Hero of the Greek Alexandrian School, we have the Roman architect-engineer Vitruvius [circa 27 AD]. He is one of the links between Greek and Roman machine designers. His work is cited as influencing Leonardo, because the books of Vitruvius were rediscovered in the fifteenth century and Francesco di Giorgio attempted to write a translation. There is also the influence of thinkers such as Roger Bacon and scholars in mathematics and mechanics in the late Middle Ages such as Villard Honnecourt or perhaps the Arab writer al-Jazari, but the direct links to the Renaissance engineers is not clear.

What is clear is the influence of Filippo Brunelleschi and his chroniclers such as Ghiberti and Sangallo, who made drawings of many of his construction machines. (see e.g. Battisti (1981, 2002, pp 132–136), Galluzzi (1997, pp. 93–116)) Another established line of influence is that of Mariano Taccola and Francesco di Giorgio Martini. (Reti 1963) Since Leonardo had a copy of one of di Giorgio's books in his library, a direct connection between Leonardo and Francesco di Giorgio seems appropriate. Leonardo also had copies of the works of other nodes of the chart, such as Leon Batista Alberti, Roberto Valturio, and Buonaccorso Ghiberti. Some of the links are speculative as to influence on Leonardo's machine designs, such as that of Giuliano da Sangallo. Earlier machine catalogs of the late fourteenth century such as those by Giacomo Fontana and Kyeser (c.1405) are likely to have influenced fifteenth century engineers but the direct evidence is not clear.

The influence lines of Leonardo himself on other contemporary and later nodes are more problematic since his manuscripts were never published. However some scholars, such as Ivor Hart (1961) believe that after Leonardo's death in 1519, the executor of the manuscripts, Francesco Melzi, may have allowed certain artists to make copies from the manuscripts or allowed scientists and scholars to read some of the manuscripts in the half century period that Melzi had them in his possession. Two possible lines of influence of Leonardo lead to Fazio Cardano [1444–1524] and his son Geronimo (Jerome) Cardano [1501–1576].

Within the half century period of Melzi's caretaking, important machine books of the Italian Ramelli (1588) and Besson (1571) were published. The machine book on mining by the German, Agricola (1556) also appeared in this period as did one on metallurgy by Biringucci (1540). According to Gille (1966) other technology books on agriculture, chemistry and arms appeared in sixteenth century after Leonardo's

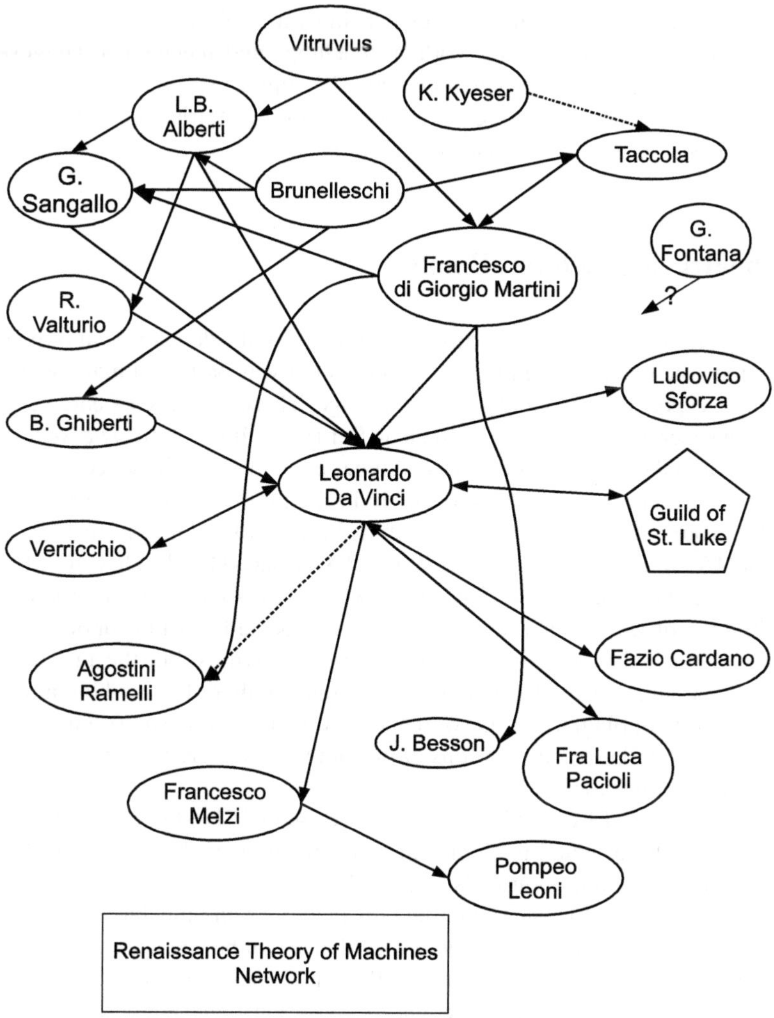

Fig. 2.3 Influence network of Leonardo da Vinci 1350–1550 (*Source* Moon (2007), Scaglia (1985, p. 92))

death by authors from what is now England, Spain, Germany and Italy. This plethora of knowledge books on engineering and technology was one of the hallmarks of the Renaissance machine age (see Table 2.2).

The Italian engineer and historian Ceccarelli (2008), has published a survey of the early machine and mechanism designers and theorists of the Renaissance and lists a dozen other nodes in the early machine network that are not included in the Leonardo Network of Fig. 2.3.

We can debate the number of nodes and links as well as the direct or indirect influence on and by Leonardo's work in machines and mechanics. However the overall features of the influence network in Fig. 2.3 support the premise that workshop knowledge of machine creation evolved over several centuries and involved many other nodes besides that of Leonardo.

2.4 Theatre of Machines Books

Contemporary historians of technology recognize that the development of machines was not the sole result of the inventor-geniuses but was part of human evolution of knowledge, especially craft and guild knowledge. These theories appeared a generation ago in several important books such as Bertrand Gille (1966). *The Renaissance Engineers* and the five-volume work of Singer et al. (1956). Evidence for this view of machine invention was always available in the published machine books of the fifteenth through eighteenth century, such as Francesco di Giorgio (c.1470–1480), Agricola (1556), Besson (1571), Ramelli (1588), Zonca (1607), De Caus (1615) de Caus, Böckler (1661), (see Fig. 2.4) and Leupold (1724), in a succession of books titled '*Theatre of Machines*' (Table 2.2). What is especially striking about these publications, is not only the artistic quality of the engravings, but the fact that some of these artist-engineers blatantly copied from one another often without attribution. Were they guilty of plagiarism? Or were they simply recording such common knowledge that their similar representations of machines were merely symbols or icons of contemporary technology?

Before the time of Leonardo's notebooks (c.1480–1515) there were many manuscripts in circulation relating to ancient, existing and new designs for machines as well as works by contemporaries (see Table 2.2).

It is natural to ask if Leonardo copied his machines and mechanisms from other machine books and manuscripts. There is evidence that he had access to the work of Konrad Kyeser and Roberto Valturio as well as the Sienese artist-engineer, Francesco di Giorgio. One of the volumes in Leonardo's library is believed to have been a book of drawings by Francesco di Giorgio similar in style to Leonardo's notebooks but presented in a more orderly format.

Two of the most influential machine books in the period after Leonardo's death were those of Besson (1571) and Agostino Ramelli. Both works exhibit beautiful machine lithographs in isometric views that provide much more detail than the sketches of machines in Leonardo's manuscripts. Each of the three authors have some machines in common leading some to speculate as to whether these later engineers had access to Leonardo's manuscripts after his death in 1519. Reti (1980a) In several examples of machines given by Ramelli and Besson, one can also find a similar machine in the work of Francesco di Giorgio whose books were widely circulated in the sixteenth century. It would be a mistake to attribute all the machine designs of the sixteenth century to fifteenth century artist-engineers however, as one can see by comparing the drawings.

Table 2.2 Theatre of Machine books

Author	Publication date or *life span*	Short title	City of publication or *author*
Ibn al-Razzaz al-Jazari	1204–1206	The book of knowledge of Ingenious mechanical devices	mesopotamia
Villard de Honnecourt	1270	The sketchbook of Villard de Honnecourt	
Guido di Vigevano	1335	Texaurus	*Pavia*
Giacomo Fontana	1393–1455	Bellicorum instrumentorum liber	Venice
Konrad Kyeser	1366–c.1405	Bellifortis	*Bavaria*
Giuliano da Sangallo	1445–1516	Opusculum	Florence
Ghiberti Bonaccorso	1451–1516	Zibaldone	Florence
Marianus Jacobus/Taccola	c.1450	De Ingeneis	Siena
Franceso di Giorgio	1470–1480	Trattato di Architettura	Siena
Roberto Valturio	1472	De re militari	Verona
Leonardo da Vinci	1480–1515	*Notebooks* (*unpublished*)	Florence, Milan
Vannoccio Biringucci	1540	De la Pirotechnia	Venice
Georgius Agricola	1556	De Re Metallica	Basel
Jacques Besson	1578	Theatre des instruments mathematiques et mechaniques	Lyon
Jean Errard	1584	Le Premier Livre des Instruments Mathematiques Mechaniques	Nancy
Agostino Ramelli	1588	Livre des diverses et artificieuses machines	Paris
Jacob de Strada	1617–1618	Künstlicher Abriss allerand Wasser, Wind, Ross und Handt Mühlen	Frankfurt
Heinrich Zeising	1613	Theatri Machinarium	Leipzig
Vittorio Zonca	1607	Novo teatro di machine et edifici	Padua
Salomon de Caus	1615	Les Raisons Des Forces Mouvantes avec diverses Machines	Frankfurt

(continued)

Table 2.2 (continued)

Author	Publication date or *life span*	Short title	City of publication or *author*
Georg Böckler	1661	Theatrum machinarum novum	Nürnberg
Otto von Guericke	1672	Experimenta nova, ut vocantur Magdeburgica	Amsterdam
Jacob Leupold	1724	Theatrum machinarum generale	Leipzig
Denis Diderot. & Jean d'Alembert	1751	Encycopedie ou Dictionnaire raisonne des sciences, des arts et de metiers	Paris
J. M. Lanz & Augustin Betancourt	1808	Analytical Essay on the Construction of Machines	Paris
J.-A. Borgnis.	1818	Traite Complet De Mecanique: Composition des Machines	Paris
Ferdinand Redtenbacher	1857	Die Bewegungs Mechanismen	Mannheim
Franz Reuleaux	1861	Der Constructeur	Berlin

Fig. 2.4 Design for a dual cylinder pump from *Theatrum Machinarum Novum* by Böckler (1661) (see Table 2.2). Note the valves at the bottom of the cylinders

2.4.1 Jacque Besson [c.1540–1573]

Jacque Besson was born in or near Grenoble, France, lived for a time in Lausanne and Geneva, Switzerland, returned to France and then fled to England in the wake of French persecution of French Protestants. According to records, Besson worked as a water pump engineer, became a Protestant minister, and then taught mathematics in Lyon and Orleans in the period 1564–1567. Besson reportedly met French King Charles IX in Orleans at this time and described to him his inventions of new machines. Besson published the first edition of his famous machine book in 1571–1572 and again in 1578 with 60 engravings and brief descriptions of the machines, under the title *Theatre des instruments mathematique et mechaniques*.

Besson's book contains six plates of mathematical drawing instruments. The remaining plates present machines for sawing lumber, mills, pumps, lathes, and construction equipment.

2.4.2 Agostino Ramelli [c.1531, c.1600]

Ramelli was born in a small village northwest of Milan and was a military engineer for French kings. As a military engineer, we would assume he had practical experience with at least some of the machines he designed in his published book. Besson on the hand, born in France, was more or less a mathematician. His unpublished manuscript that preceded his published book contains so-called 'principles' for design of machines. It is possible that many of Besson's machines were drawn from existing books or from his imagination.

Ramelli's 1588 work *Livre des diverses et artificieuses machines* has been translated and printed by Dover Publications in 1987, edited by Gnudi and Ferguson. There is a wealth of scholarship in this edition, including an inventory of types of machines, and machine components in the nearly 200 plates.

Although Ramelli worked as a military engineer, more that half the plates in his book concern water-raising machines and only about two dozen plates deal with military machines. It is likely that some devices were drawn to represent contemporary machines while others were invented in the imagination of Ramelli. Like Francesco di Giorgio, we can view many of these drawings as an exploration of the design space, for instance, when he shows three plates on toroidal water pumps.

In a paper, written in 1972 after the discovery of Leonardo's Madrid Codices in 1967, the Leonardo scholar Ladislao Reti published a detailed analysis of the machines of Ramelli and Leonardo da Vinci. He gave over ten examples of nearly similar machine or kinematic mechanism designs appearing in both Ramelli's book and Leonardo's manuscripts. Acknowledging the fact that Leonardo's machines were similar or influenced by the earlier machine books of the fifteenth century, Reti nonetheless argued that Ramelli must have had some access to the Leonardo manuscripts after his death in 1519 or that knowledge of the work of Leonardo was so widespread that it became part of the common wisdom and practice of the machine craftsmen of the time. However Reti gave no specific instance in which Ramelli would have had access to either the owner of the manuscripts, Francesco Melzi, or the manuscripts themselves.

2.5 Evolution of Machine Engineering

In his Notebooks, Leonardo used the terms *invention*, *design* and *engineer* many times, providing evidence that such concepts were in common use in the Renaissance. His famous letter to the Duke of Milan is full of bragging about his engineering prowess. Only at the end of this long letter does he mention his skills as an artist and painter. Modern historians sometimes describe these men as artist-engineers, or architect-engineers such as Taccolo, Francesco di Giorgio, or Brunelleschi since their skills in design were often commensurate with their ability to draw, copy and sketch. These artist-engineers often had a good knowledge of geometry. Thus one can

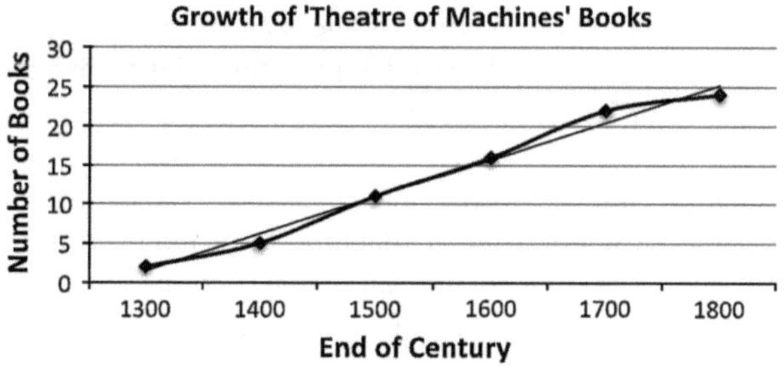

Fig. 2.5 The linear growth of 'Theatre of machines" books from the fourteenth to the nineteenth centuries based on Table 2.2. The straight line is a 'best fit' using Excel software

certainly find the seeds of a professional class of machine designers in the Renaissance, similar to architects, who are distinct from machine *builders* or craftsman.

Another element of the evolution of machine engineering is the notion of *modular elements*, or the concept of what Reuleaux called *constructive elements* of machine design. Reti (1980a, b) has been previously cited for comparing the drawings of machine elements and mechanisms in Leonardo's *Codex Madrid* with the Reuleaux's basic list of 22 *constructive elements.* In the *Codex Madrid*, one can identify modular elements such as gear pairs, pumps, cranked linkages, endless screws, pulley systems, ratchets, bearings, and escapements which went beyond the obligatory list of Aristotelian simple machines of the wheel, lever, pulley and inclined plane. Similar complex machines are found in many machine picture books before and after Leonardo, such as those by Roberto Valturio, Francesco di Giorgio, or Jacques Besson that indicates modularization in machine design was solidly under way by the early fifteenth century. Following the idea of an avalanche model for invention, we can argue that the plethora of machine elements existing at the time of Leonardo provided the critical mass of ideas that enabled him to propose so many new machines.

At some stage in this machine evolution the slow diffusion of knowledge was augmented with the shock wave of a new invention, i.e. a major departure from the conventional practice. The avalanche theory of invention is such that a slow period of improvements and materials development brings together opportunities that erupt in a spurt of new inventions as we shall illustrate below in the invention of precision clocks. Certainly the case for this model can be made in the early days of the Watt-Boulton steam engine circa 1790. A more contemporary example is the Internet and communications revolution of the late 1990s. Is there evidence that an avalanche development occurred in the Renaissance era of Leonardo da Vinci?

Although there was a steady advance in the theory and practice of designing and building machines beginning in the fifteenth century and progressing into the seventeenth and eighteenth centuries, there is no strong mathematical evidence for an avalanche of knowledge as we saw in the steam engine at the end of the eighteenth century. One measure of the advance of machine knowledge is the accumulated number of theatre of machines books as shown in Fig. 2.5. Here we see a linear growth in time and not the sharp rise one would expect from an avalanche of knowledge. This may be explained first by the lack of an efficient communication system and transportation system during the three centuries from Leonardo to James Watt. Second the lack of key technologies such as strong materials and fabrication methods may have impeded an earlier industrial age. And there weren't the right political tools such as patent systems and economic systems in place. One might say however that the slow build-up of machine knowledge from the Renaissance to the eighteenth century provided the 'knowledge commons' that eventually led to the exponential rise of the steam engine and other technologies of the industrial revolution.

2.6 Kinematics of Machines and Franz Reuleaux

Franz Reuleaux [1829–1905] was well known for his theories in machine engineering during the late Industrial Age of the nineteenth century . He was also a player on the stage of new technologies. He was a principal consultant to Otto and Langen on the gas combustion engine (see Chap. 3) as well as a friend of Siemens. He was ambassador to international expositions, a proponent of German educational theories and an advocate for machine style and art in the industrial age. Reuleaux was not an inventor like James Watt, or an entrepreneur like Werner Siemens, nor was he a scientist like Faraday or Henry. Yet he played crucial parts in the later stages of the nineteenth century machine age, in roles of engineer-scientist, professor, university head, and advisor to industry and government. Reuleaux was an important node in several technology networks. (Zopke 1896–1897)

Franz Reuleaux was the teacher of Lilienthal, the world's leading glider designer and pilot in the decade before the Wright brothers became interested in flying machines. (see Chap. 4) Reuleaux appears as a node in both the internal combustion history network (Chap. 3) and the early aviation network. (Chap. 4) His main recognition however is as one of the principal nodes in the network for the theory of kinematics of machines.

In 1864 Reuleaux began his leadership positions in Berlin, where he would spend the rest of his career first as director of the Gewerbeakademie [1868–1879] and then as rector [1890/91] of the combined Gewerbeakademie and the Bauakademie that formed the Technische Hochschule Berlin. At TH Berlin, Reuleaux was the director of mechanical engineering. He also held the title Royal Privy Councilor, and served as a consultant to the new German Reich. Reuleaux was a member of the Imperial Patent Office for 8 years (Fig. 2.6).

After leaving Karlsruhe, Reuleaux published a handbook on machine design in 1854 with a fellow student Carl L. Moll. Reuleaux's professor Ferdinand Redtenbacher at Karlsruhe was also a teacher of Karl Benz the inventor of the automobile. Reuleaux published the first edition of his popular machine design text *Der Constructeur* in 1861. This work went to four editions and appeared in four languages, including an English edition in 1893. This text contains descriptions of many different types of machines as well as machine components. The first editions of this

Fig. 2.6 German Engineer Franz Reuleaux, consultant to Otto and his internal combustion engine, professor to Otto Lilienthal of glider fame and international traveler establishing links between European machine engineers and North America. He was one of the principal theorists in the kinematics of mechanisms

work had very little kinematics of machines. Reuleaux's German text *Theoretische Kinematik* appeared in 1875 and was quickly translated into English in 1876 as *The Kinematics of Machinery: Outlines of a Theory of Machines*. A second volume appeared in German in 1900.

2.6.1 'Father' of Kinematics of Machines

Unlike James Watt, who was an instrument maker and craftsman, Reuleaux and his fellow engineer-scientists were trained in science and mathematics, philosophy and literature as well as in "mechanical arts", influenced in part by the French 'Polytechnique' tradition with its strong emphasis on mathematics and mechanics. Unlike the craftsman-engineer who believed in trial and error and hands on education, the engineer-scientist believed that machines could be created and designed using scientific principles guided by rigorous mathematics. This period was the beginning of the importance of universities as key nodes in the innovation networks of the late industrial revolution. In the kinematics of machines, the principal university nodes were Ecole Polytechnique in Paris of Gaspar Monge , Cambridge University of Robert Willis (1841, 1870) and the Berlin Technische Hochschule of Franz Reuleaux.

Franz Reuleaux's major work in kinematics was translated almost immediately into English by Professor Alexander B.W. Kennedy of University College London in

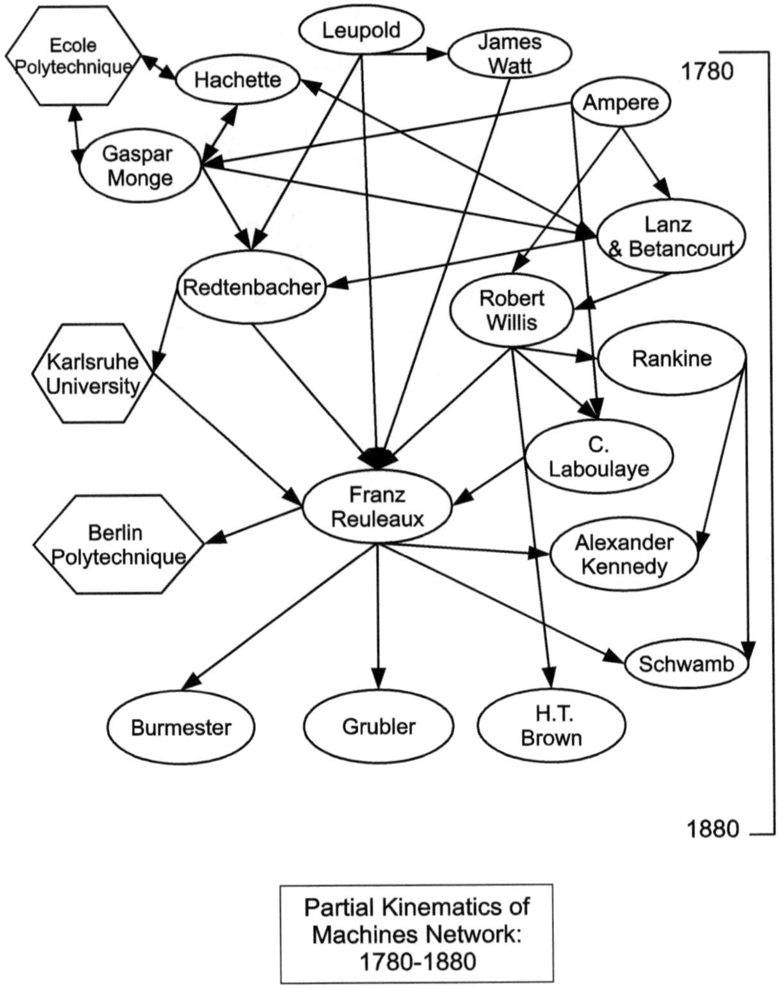

Fig. 2.7 Influence network of Franz Reuleaux in the nineteenth century related to the kinematics theory of machines (Moon 2007)

1876 under the title, *The Kinematics of Machinery: Outlines of a Theory of Machines.* It was also translated into French and Italian, such was its impact on the engineering community in Europe. The ideas that Reuleaux presented in this book influenced the field of machine design for a century. A graphical summary of Reuleaux's influence network is shown in Fig. 2.7. His theory of machines was seen as genius by many of his contemporaries and early generations of kinematics theorists into the twentieth century. He was the first to introduce the ideas of topology into the theory of machine mechanisms (Fig. 2.8).

Fig. 2.8 Models of kinematic pump mechanisms created by Franz Reuleaux to illustrate his kinematic theory of machines (Models from the Cornell University Collection of Reuleaux Kinematic Models, Ithaca, NY. (Moon 2004) Go online and search for "KMODDL" for photos and videos)

Reuleaux introduced his *symbol notation*, a language of machines, which extended the ideas of Charles Babbage, one of the pioneers of computer science. His sequence of kinematic constraints in a mechanism became a sequence of symbols, each representing a unique geometric constraint. Reuleaux's principal philosophical question in his theory of machines was; 'How did the machine and its mechanisms come into the mind of the inventor?' In the Introduction to his book he wrote,

"*What is left unanswered is however the other, immensely deeper part of the problem, the question: How did the mechanism, or the elements of which it is composed originate? What laws govern its building up? Is it indeed formed according to any laws whatsoever?*"

To support his theories of machines, Reuleaux built a collection of 800 kinematic models in Berlin and sold a subset of 350 models to universities as aids to teaching his theory. In 1876, a few hundred models were exhibited in London and excited Professor Kennedy at University College London who translated Reuleaux's book on kinematics of mechanisms into English, a year after its German publication. Reuleaux's network consisted not only of university professors of mechanical engineering, but also publishers, politicians, practicing engineers and businessmen. These latter nodes are not shown in Fig. 2.7, but some appear in the internal combustion engine network discussed in Chap. 4.

In addition to having major industrialists such as Siemens (telegraph and electric power) and Langen (Otto's partner) as friends, Reuleaux produced some famous students such as Lilienthal of glider fame, Mannesman (seamless pipe), Linde (cryogenics, air conditioning) and Junkers who later became a major airplane producer.

Reuleaux can also take some credit for driving a student to fame outside of engineering, namely the American photographer Alfred Stieglitz. Sent by his New York father to study mechanical engineering with the famous Berlin engineer, Stieglitz found Reuleaux's lectures so boring and that he quit engineering and took up the study of photography.

Reuleaux and his generation saw the Age of the Machine as a network of progress reaching back to the Greeks and Egyptians as part of the destiny of humankind. Machines were the embodiment of man's knowledge and control over nature. He viewed the evolution in the development of the machine as an analog to the development of advanced societies in which education, crafts, manufacture and government are linked in a chain of mutual dependency for the common good (Reuleaux 1885).

A comparison of Leonardo's design ideas with those of Reuleaux in the mid nineteenth century is a leap forward through 400 years of machine invention and development especially noting the eighteenth century work of Newcomen and Watt and the steam engine. Reuleaux's books on machine design are a good place to see the end point of the evolution of machine construction from the workshop model in England to formal mechanical engineering training in the Germanic countries, or what in German is called *Maschinenbau* or 'machine building'. In Reuleaux's books we see explicit use of mathematics, experimentally measured properties of materials for machine construction, summaries of best practice and modular machine elements and mechanisms.

Lest one think that Reuleaux fostered a dull, rote order of machine design, one can find in his *Kinematicsof Machinery* 1876 a passionate plea for a study of synthesis and 'invention' and an admission of the failure of machine theory to come up with a process of synthesis in machine design. '*How did Watt invent the straight-line mechanism in the steam engine?*', Reuleaux asked rhetorically.

It is clear from his writings that Reuleaux viewed the development of new machines as one of evolution, and that every invention has had an antecedent developed further by clever inventors using new scientific discoveries as well as the marketplace. (Kinematics of Machinery 1876, p. 231);

"Very gradually each invention came to be used for more and more purposes than those for which it was originally intended, and the standard by which its excellence and usefulness were judged was gradually raised. —These attempts resulted in further improvements and these in turn led once more to new applications and more extended use."

Reuleaux's general interest in the history of invention is exhibited in an eight volume series that he served as editor, *The Book of Inventions, (Buch der Erfindungen)* a pictorial, popular book on the history of invention from the early Egyptians to the end of the nineteenth century (Reuleaux 1884). He did not accept the contemporary theory of invention as resulting from scientific discovery, a view that is often expressed in popular literature on technology in the United States. Nor did he believe in the discontinuous genius theory of invention, where the 'hero' inventor, working alone, makes an important advance that benefits humankind. He considered both scientific discovery and technical invention as evolving from a tension between the two, sometimes within the same man.

2.7 The Evolution of Clock Technology Networks

As an early example of the avalanche theory of machine invention, we examine the development of precision clocks in the eighteenth and early nineteenth centuries. One of the important tools for modern science were instruments for measuring time. From the thirteenth to the twentieth century precision instruments for measuring time were mechanical assemblages of dynamic and kinematic elements that we call clocks or chronometers. Mechanical technology replaced the water, sun and sand clocks in use prior to the thirteenth–fourteenth centuries. Moreover the first important mechanical clock element was the escapement that converted steady potential energy of gravity into oscillating kinetic energy of a time increment such as the verge and foliot escapement used in many cathedral clocks. The discovery of the dynamic principles of the pendulum by Galileo and others in the sixteenth century, and its use in the clock as a regulator by Christiaan Huygens in 1657 heralded the beginnings of modern experimental science. These early clocks embodied the basic elements of modern digital computers in that they incorporated a digital time measuring system (escapement), time regulator (pendulum or balance wheel and spring), information storage in the striking mechanism in the form of a notched wheel and an information retrieval system coupling the time keeping and striking mechanisms.

Mechanical clocks also embraced the advancing field of kinematic mechanisms employing geared wheels, cams, levers, and many other machine elements that were slowly being cataloged in the works of Leonardo da Vinci and the later writers of the 'Theatre of Machines' books of the fifteenth–eighteenth centuries (Fig. 2.9).

From many historical sources, we can assemble a timeline of both clock technology advances as well as the people and association nodes that constituted its evolutionary history.

A short timeline event list is as follows: (Landes (1983) *Revolution in Time);*

1656: *Christiaan Huygens*; [1629–1695]; built first clock with pendulum regulator; wrote *Horologium* 1658
1666: *Robert Hooke* [1635–1703] and *William Clement*; anchor escapement;
1675: *Thomas Tompion* [1639–1713]; cylinder escapement, long case clocks;
1676: *John Flamsteed*; Royal astronomer; lunar tables for longitude;
1714: *Board of Longitude*, London; Prize of 20,000 Pounds;
1720's: *Henry Sulley*; jewel bearings to reduce friction;
1715: *George Graham* [1673–1751]; deadbeat escapement;
1762: *Nevil Maskelyne*; royal astronomer, longitude method of lunars;
1754: *Thomas Mudge* [1715–1794]; lever escapement;
1735–1759: *John Harrison* [1693–1776]; James Harrison and William Harrison, first precision maritime chronometers; winners of 'Longitude Prize';

Fig. 2.9 Compound pendulum gravity escapement model from the Reuleaux kinematic models collection, Cornell University. A version of this escapement was used in the tower clock of Westminster in London. Thomas Mudge [c.1774] is credited with making a reliable gravity escapement

1760: *Ferdinand Berthoud* [1727–1799]; maritime sea clocks, *Traite des horloges marines;*
1748: *Pierre Le Roy* [1717–1785]; detent escapement, compensation balance;
1784: *Thomas Earnshaw* [1749–1829]; self compensating balance;
1773: *John Arnold* [1736–1799]; self compensating balance;
1782: *Abraham-Louis Breguet* [1747–1823]; self winding watch;
1814: *John Dent* [1790–1853]; watch maker to Napoleon, designer of Big Ben.

A partial historical innovation network, on the development of precision clock devices suitable for determining longitude from ships is presented below in Fig. 2.10. The historical period spans from 1650 to 1800. The England bias reflects the clock history sources.

One of the important components of the exponential rise in technologies at the dawn of the Industrial Revolution is the talents of skilled artisans and systems to train them. By the beginning of the fifteenth–sixteenth centuries there were many horology guilds in Europe with hundreds of members. Thus by the beginning of the eighteenth century the number of horology artisan nodes was already large. Some of these guilds are listed below with their founding date.

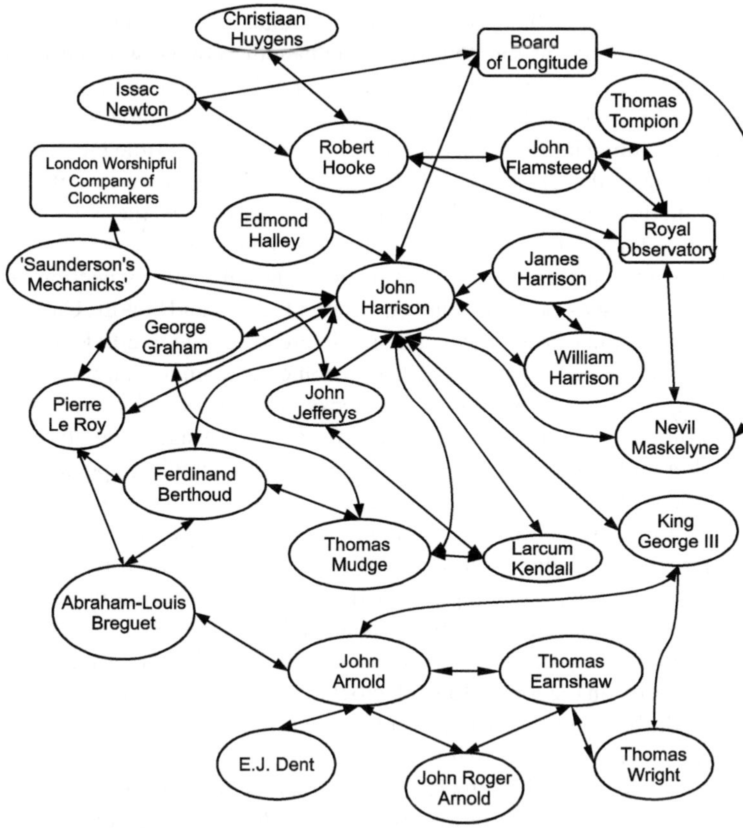

Fig. 2.10 Partial historical clock network that led to the emergence of exact maritime time-keeping to determine longitude in the period 1650–1800 based on Sobel (1995), Landes (1983) and Bruton (1967)

Bloir French Clockmakers Guild; 1515
Nuremberg Clockmakers Guild; 1565
Paris Clockmakers Guild; 1544
The London Worshipful Company of Clockmakers; 1632
Geneva Clockmakers; 1600
Black Forest Clockmakers; 1670

Key horological names who have been recognized by historians as principal nodes of the network of clock greats include John Harrison, Pierre Le Roy , Breguet, Lord Grimthorpe etc. If one takes these clock nodes from a recognized historian such as Eric Bruton and others, one can create a graph illustrating the growth of the number of clock related nodes versus time as shown in Fig. 2.11. Surprisingly this curve shows the classic exponential rise that is characteristic of other technologies a century later. Part of the rise of nodes at the end of the eighteenth century was the interest in marine chronometry that competing countries such as England and France depended on for trade and military superiority. The popular book *Longitude* by Dava Sobel (1995) describes John Harrison's quest for the accurate marine time piece motivated by prize money from the British Board of Longitude. This book gives a glimpse into the future of the lure of money and prestigious medals such as the Nobel Prizes in focusing many efforts in the advance of technology and science.

The clock historical social network also shows the intimate connection between the work of clock artisans, scientists and government agents and boards. Clock history in the nineteenth century brought in economic interests related to efficient and cheap manufacturing, as well as artistic style, that become more important than the nature of the mechanical mechanisms inside the clocks and watches. Another interesting aspect of horological history is the fact that over several design generations, clocks became smaller and more complex, from cathedral tower clocks to pocket watches, and the accuracy of clocks became greater, a trend we have seen paralleled in the development of modern digital computers in the twentieth century.

It should be acknowledged that the construction of innovation timelines, influence diagrams, social network diagrams and growth of innovation nodes graphs is sometimes dependent on the ethnic background and language of the historian whose narrative is used for the data. Thus the books consulted in this study were written in English and focused on the role of English clockmakers. If one used French historical texts there might be other nodes that would appear that are not included in the rise of clock makers social network. However it is asserted that the exponential rise in the number of nodes related to clocks would remain similar to that in Fig. 2.11. In all of the technologies studied in this book, from clocks to engines to radios, the historical networks were international in nature. Expertise in clock-making moved from England to France to Germany and later to America in the nineteenth century.

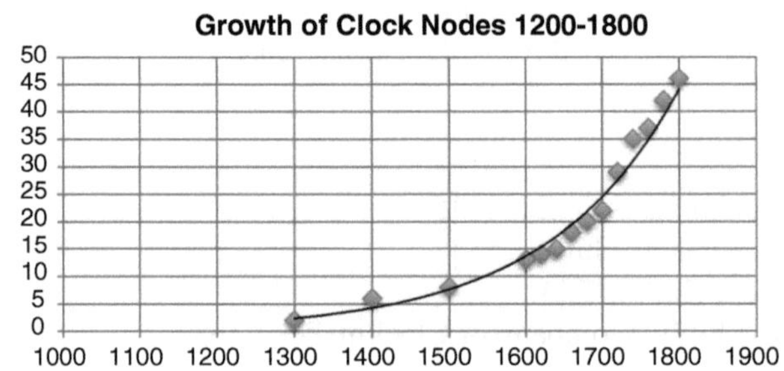

Fig. 2.11 Exponential growth in the number of human nodes in the clock history network based on Bruton (1967). The solid line is a 'best fit' exponential curve using EXCEL software

2.8 Growth of Kinematics of Machines Nodes: Ferguson's Timeline

One can see over and over again in the history of technical and scientific ideas, the observation of exponential growth in the number of historical players in a new field. The sum of nodes and events in the history of a technical field contributes to the pool of common practices, or knowledge commons, that inspires others to enter the network . This is true in the history of kinematics of machines network shown in Fig. 2.12 below. This figure is based on the narrative description of the history of kinematics by the well known scholar Eugene Ferguson (1962), once a curator at the Smithsonian Museum in Washington DC. As noted earlier there was a steady progress in the theory of machines as evidenced by the growth of the number of published "*theatre of machines*" books from the fifteenth to the early nineteenth century. (Fig. 2.5) While the earlier growth in kinematics was approximately linear in time, the exponential growth in kinematics of machines in the second half of the nineteenth century occurred when there was a concomitant growth in communication and transportation networks in Europe and North America. We shall not be surprised by a similar progress in internal combustion engines and aviation in the late nineteenth century as discussed in Chaps. 3 and 4.

The 'Age of Machines' spanned Watt's remarkable improvements to the steam engine in the last quarter of the eighteenth century to the Wrights' development of powered flight at the beginning of the twentieth century. In the early nineteenth century, machine technology was a workshop process passed on to apprentices by master mechanics and engineers who often kept their methods secret and guarded against use by their competitors. The steam engine not only sparked a revolution in the creation of a mobile energy source, but also resulted in the evolution of new methods of creating machines. The wresting of machine design from the workshop began in the Ecole Polytechnique in Paris in the late eighteenth century with the work of Monge and Hachette, and later by Ampere and Lanz and Betancourt. These ideas were further developed in Britain, especially in the work of Robert Willis [1800–1875] and

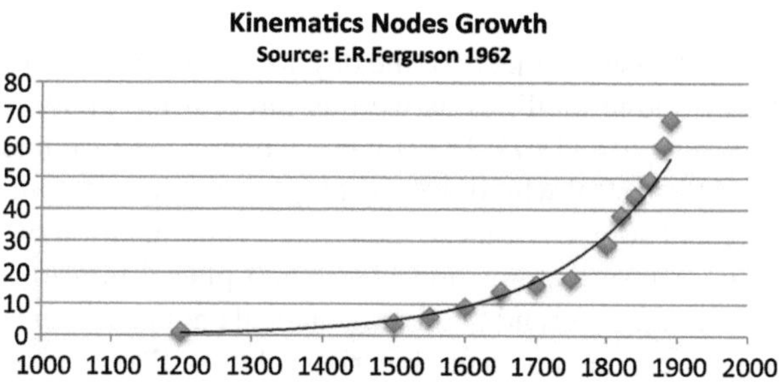

Fig. 2.12 Exponential growth in the number of historical nodes in the history of kinematics of machines (Based on Ferguson (1962)). The solid curve is a 'best fit' exponential curve using the software package Excel

William Rankine [1820–1872] and in Germany by Ferdinand Redtenbacher [1809–1863] of the Polytechnic School at Karlsruhe whose student was Franz Reuleaux [1829–1905].

2.9 Networks of Machine Designers and Builders in the Nineteenth Century

The workshop system in Britain during the early machine age was characterized by a close relationship between master mechanics and young engineers through apprenticeships that generated a social network of machine builders. For example, Henry Maudslay [1771–1831] trained with the lock manufacturer Joseph Bramah [1749–1814] who invented the hydraulic press. Maudslay also worked with the great civil engineer Sir Marc I. Brunel [1769–1849]. Brunel had visited the American engineer John Stevens. Maudslay later trained engineers and toolmakers Joseph Whitworth [1803–1887], James Nasmyth [1808–1890], and Joseph Clement [b. 1779]). Clement was hired in the 1820s to build Charles Babbage's famous kinematic calculating machine. What all of these great engineers had in common was a lack of formal engineering training. This training of machine designers and builders was not too different from the training of the Renaissance engineers such as Mariano Taccola, Francesco di Giorgio, Fillippo Brunelleschi and Leonardo da Vinci.

On the other hand the nineteenth century machine theorist Robert Willis [1800–1875], educated in the mathematics tripos at Cambridge University, clearly came from a different pedigree than most engineers of his generation (see Moon (2003b), for a short biography of Robert Willis). Willis was elected as a Fellow of the Royal Society in 1830, and taught at Cambridge at the same time as the mathematician Charles Babbage, who designed the forerunner of the computer. Willis made drawings in his personal sketchbook of Babbage's famous 'difference machine' calculator. Franz Reuleaux came from a family engineering workshop tradition in Belgium that made machines and pumps. He broke that mold by obtaining his engineering education at the universities at Karlsruhe, Berlin and Bonn based on mathematics, science and philosophy.

Among the great British engineers of the day were I. K. Brunel [1806–1859] who built major rail and bridge facilities as well as three of the greatest steamships of the time, Joseph Whitworth [1803–1887] who standardized machine elements, and James Nasmyth [1808–1890] who developed high precision machining techniques. Yet in the biographical sources of these men, there is no mention of any interaction with Robert Willis, whose lectures and books had placed him in the forefront of engineering theory. In England and Scotland practical engineers were trained through an apprenticeship system and not in engineering schools. Brunel's famous father, Sir Marc I. Brunel tried to groom his son for admission to the l'Ecole Polytechnique in Paris, but Isambard K. did not pass the competitive exam and instead served his apprenticeship with a famous horologist, Louis Breguet, in Paris. The contrast with the situation in Germany a generation later could not have been greater.

Reuleaux received formal education at the Polytechnischen Schule at Karlsruhe under Ferdinand Redtenbacher, as well as practical training in his uncle's company. Carl Benz [1844–1929] also received his training from Redtenbacher, as did Eugen Langen who with Nicholas Otto developed the internal combustion engine. Langen and Reuleaux were classmates at Karlsruhe and Reuleaux played a pivotal role in the early development of the Langen-Otto enterprise. Redtenbacher is known as a pioneer in Germany's development of mechanical engineering education. (Wauer et al. 2010) By the end of the century, the route of professional education of many famous German engineers started in the university with a background of mathematics and engineering science as well as practical training.

2.9.1 International Expositions

In the network of machine builders, another set of nodes that facilitated technical communication in the nineteenth century was the international exhibitions that attracted millions of visitors. The first major event was at London's Crystal Palace in 1851. Reuleaux's network was aided by his travel to World Exhibitions in London [1862], Paris [1867], Vienna [1873], Philadelphia [1876], Sidney [1879], Melbourne [1881] and Chicago [1893]. His professional life coincided with new communications networks such as overseas mail and the telegraph that linked the growing industrial world with the first Internet. Reuleaux played a major role in many of these 'world's fairs' for over 40 years often as Germany's ambassador.

Reuleaux was a player in the political network of the machine age. One of his most famous roles was as the German ambassador to the Centennial Exhibition in Philadelphia in 1876. In his official duties at Philadelphia he sat on judging panels and wrote articles on industrial advances exhibited at the Exposition. Later at the 1893 Columbian Exposition in Chicago, Reuleaux was questioned by reporters from several technical magazines as to whether German manufacturers had improved their goods over his "*billig und schlecht*" description of 1876. He proudly responded that German goods were quite good but that the Americans had continued a lead in precision manufacturing, to which he was widely quoted in the American press and again criticized at home. Reuleaux's reputation in North America was such that he was awarded "honorary member" of the American Society of Mechanical Engineers.

Unlike the theme park atmosphere of twentieth century World's Fairs, these nineteenth century extravaganzas provided an opportunity for countries to show off new machines and technology that attracted both the average public as well as technical experts. For example, Henry Ford's father visited the 1876 World's Fair in Philadelphia and told his young son about the wonders he had seen there. Otto had sent an engineer to exhibit his new internal combustion engine at the Philadelphia Fair of 1776. There is no doubt that Reuleaux and other engineers used these occasions to

Fig. 2.13 The world's largest steam engine: the Corliss exposition engine at the 1876 Philadelphia centennial exposition. Its two cylinders produced 1400 HP and powered the exhibits in Machinery Hall. Franz Reuleaux was the German Ambassador to the centennial exposition. By 1910 this engine was sold for scrap, made obsolete by the steam turbine and the internal combustion engine

learn about technology in other countries, which is reflected in the comments and footnotes in his books (Fig. 2.13).

2.10 Cheering for Leonardo

Studies of the machines of Leonardo da Vinci have had at least three camps; (i) those who want to see him as a lone genius and view his designs as prescient for our current technology, (ii) those critics who claim his drawings simply reflected the technology of his times and earlier machine books and (iii) a few who think Leonardo was the best of a network of machine designers of his day. The scholarly work on da Vinci's painting and art by Kenneth Clark certainly falls in the cheerleader group as does Reti. In the critics group we have Duhem and Truesdell (1968) and perhaps Gille. Straddling this group is the work of Galluzzi (1997) of the Institute and Museum of the History of Science in Florence, whose recent work on the Renaissance engineers of Siena highlighted the work of Francesco di Giorgio. The author's book Moon (2007) has tried to steer this middle course with the thesis that in an evolutionary process, many paths and links are necessary for transmittal of technical knowledge including the genius of both the scientist-mathematician or engineer-architect as well as the genius of ordinary skilled craftspeople making incremental improvements one machine at a time.

Linking Leonardo's artistry of machines and Reuleaux's theory of mechanisms is an unbroken evolutionary network of designing and building machines, originating in the ancient civilizations, codified by the Greeks and transmitted to the

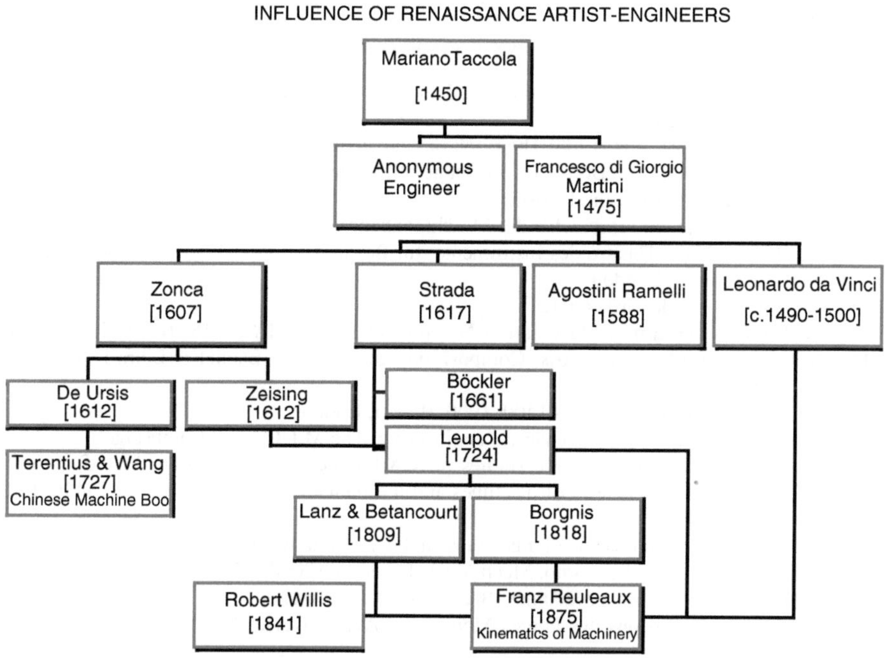

Fig. 2.14 Chart of influence of renaissance engineers on nineteenth century kinematic theory of machines engineers (After Reti 1963)

emerging European Renaissance culture by the Arab and Moslem civilization of the Middle Ages. (Fig. 2.14) Both Leonardo and Reuleaux made considerable contributions to this machine design tradition. They were both the recipients of a technical culture based on both written and workshop knowledge systems. These global technical networks stretched largely from Europe to Asia in the Renaissance and from North America to Europe in the Industrial Age. Although the specific machines and technologies were important, we have emphasized the historic role of the development of a methodology for inventing, designing and producing machine technology. For both Leonardo and Reuleaux the skill of precision machine drawing was of great importance. The use of geometry and kinematics or the 'geometry of motion' as well as algebraic relationships, were essential tools of the machine designer.

Contrary to the romanticized image of machine creation, most machine advances evolved from earlier concepts and were motivated by economic needs and opportunities. Leonardo and Reuleaux both acted under similar pressures and incentives, textile manufacturing and warfare in Renaissance Italy, and transportation and manufacturing production in the nineteenth century. In transportation technology Reuleaux was to play a secondary, but important, link in both the internal combustion network (Chap. 3) and the aviation network (Chap. 4).

References

Agricola G (1556) De Re Metallica. Hoover HC and Hoover LH (1912) De Re Metallica (trans). Dover Publication, New York (1950)
Al-Jazari, Ibn al-Razzaz (1204–1206) The Book of Knowledge of Ingeneous Mechanical Devices. Donald R Hill D (1974) The Book of Knowledge of Ingeneous Mechanical Devices (Transl and Annotated). Reidel Publication, Boston
Beck T (1899) Beiträge zur Geschichte des Maschinenbaues. Julius Springer, Berlin
Besson, J. (1571, 1578), Theatre des Instruments. In: Bevan T (ed) (1939, 1956) The Theory of Machines. Longmans, London
Biringucci V (1540) De la Pirotechnia
Böckler GA (1661) Theatrum Machinarum Novum. In: Borgnis JA (1818) (ed)Traite Complet De Mechanique Applique aux Arts: Composition des Machines Bachelier Librarie. Quai des Augustins, Paris
Bruton E (1967) Clocks and watches 1400–1900. Barker, London
Burstall AF (1965) A history of mechanical engineering. The M.I.T Press, Cambridge
Ceccarelli M (2000) Italian kinematic studies in XIXth century. In: Ceccarelli M (ed) International symposium on the history of machines and mechanisms. Kluwer Academic Publication, Dordrecht, pp 197–206
Ceccarelli M (2008) Renaissance of machines in Italy: from Brunelleschi to Galileo through Francesco di Giorgio and Leonardo. Mech Mach Theory 43(2008):1530–1542
Cianchi M (c.1995) Leonardo's Machines, Becocci Editore, Florence
De Caus S (1615) Le Raisons des Forces Mouvantes, Avec Diverses Machines Tant Utiles Que Plaisantes, Frankfurt
Diderot D, D'Alembert J (1751–72) Encyclopedia, Paris
Feldhaus FM (1922) Leonardo, der Techniker und Erfinder. E. Diederichs, Jena
Ferguson ES (1962) Kinematics of mechanisms from the time of watt, United States National museum bulletin, 228, paper 27, 185–230. Smithsonian Institution, Washington
Ferguson ES (1992) Engineering and the mind's eye. MIT Press, Cambridge
Fontana G (1393–1455) Bellicorum instrumentorum liber, Venice, Italy
Francesco di Giorgio M (c.1450) Tratatto di Architectura (Facsimile: Architettura Ingegneria e Arte Militare: Codici Torinese Saluzziano 148, Edited by Corrado Maltese, Transcription, Livia M. Degrassi, Edizioni Il Polifilo, Milano, 1967)
Galluzzi P (1997) Mechanical marvels: invention in the age of Leonardo. Guinti Press, Florence
Gibbs-Smith CH (1978) The inventions of Leonardo da Vinci. Phaidon Press, Oxford
Gille B (1966) The renaissance engineers. Lund Humphries, London
Grothe H (1874) Leonardo da Vinci als Ingenieur und Philosoph. Ein Beitrage zur Geschichte der Technik und der Inductiven Wissenschaften Nicolaische Verlag- Buchhandlung, Berlin
Hachette JNP (1811) Traité Elementaire des machines. klostermann, Paris
Hart IB (1925) The mechanical investigations of Leonardo da Vinci. Chapman and Hall Ltd, London
Hart IB (1961) The world of Leonardo da Vinci. The Viking Press, New York
Hartenberg RS, Denavit J (1964) Kinematic synthesis of linkages. McGraw-Hill Book Co, New York
Huygens C (1658) Horologium; See also English translation of Richard J Blackwell; Christiaan Huygens' The Pendulum Clock or Geometrical Demonstrations Concerning the Motion of Pendula as Applied to Clocks
Kennedy ABW (1881) The kinematics of machinery: two lectures relating to Reuleaux methods. Delivered at the South Kensington museum with an introduction by professor RH Thurston, Reprinted from Van Nostrand's Magazine. D Van Nostrand, Publisher, New York
Kennedy ABW (1876) The Berlin Kinematic Models. Engineering 22:239–240
KMODDL (2004) Kinematic models for design digital library. http://kmoddl.library.cornell.edu

Koetsier T (2000) Mechanism and machine science: its history and its identity. In: Ceccarelli M (ed) International symposium on history of machines and mechanisms, proceedings HMM 2000, Kluwer Academic Publishers, Dordrecht, pp 5–24

Kyeser K (c.1405) Bellifortis, (Facsimile, Edited by the Georg-Agricola Gesellschaft zur Forderung der Geschicte der Naturwissenschaft und der Technik, (1967) VDI Verlag, Dusseldorf, Germany)

Landes, D.S. (1983). Revolution in time, Harvard Univ. Press, Cambridge, MA

Lanz PL, de Bétancourt A (1808) Analytical essay on the construction of machines, (trans: from the French, Ackermann R). Paris, London

Leonardo da Vinci (c.1500a) Le Manuscrit A-M, Institut de France (Facsimile Edition: Les manuscripts de Leonard de Vinci: le manuscit A-[M] de la Bibliothéque de l'Institut [de France] translation in French, by C. Ravaisson-Mollien, 1881–1901)

Leonardo da Vinci (c.1500b) Codex Atlanticus, Milan (Facsimile Edition, Il Codice Atlantico della Biblioteca Ambrosiana di Milano/ Leonardo da Vinci, translation into Italian by A. Marinoni, Preface by C. Pedretti,2000, Giunti, Firenze, 3 vols.) (Facsimile Edition: published by Regis Accademia dei Lincei, 1894–1904)

Leonardo da Vinci (c.1500c) Codex Madrid or Tratado de Estatica y Mechanica en Italiano, 1493, Madrid (Facsimile Edition: The Madrid Codices, National Library Madrid, No. 8937, Translated by Ladilao Reti, McGraw-Hill Book Co., 5 vols)

Leupold J (1724) Theatrum Machinarum Generale. Christoph Bundel, Leipzig

MacCurdy E (1938) The notebooks of Leonardo da Vinci. Reynal and Hitchcock, New York

Moll CL, Reuleaux F (1854) Constructionslehre für den Maschinenbau, (Design for Mechanical Engineering) Druck und Verlag von Friedrich Vieweg und Sohn, Braunschweig

Monge G (1795) Géométrie Descriptive, Paris

Moon FC (2003a) Franz Reuleaux; Contributions to 19th century kinematics and history of machines. Appl Mech Rev 56(2):261–285 (ASME, New York)

Moon FC (2003b) Robert Willis and Franz Reuleaux: pioneers in the theory of machines. Notes Rec R Soc 57(2):209–230 (London)

Moon FC (2004) The Reuleaux models: creating an international digital library of kinematics history. In: Ceccarelli M (ed) International symposium on history of machines and mechanisms, Proceedings of HMM2004, Kluwer Academic Publisher, Dordrecht

Moon FC (2005) The kinematic mechanisms of Francesco di Giorgio Martini and Franz Reuleaux. Proceedings workshop of international Federation for the theory of machines and mechanisms; history of machines and mechanisms, Bauman Moscow State Technical University, May 2005

Moon FC (2007) The Machines of Leonardo da Vinci and Franz Reuleaux. Springer, New York

Paccoli FL (1509) Da Divina Proportione, (Part I, p 32, II, P 15, 48)

Parsons WB (1939, 1968) Engineers and engineering in the renaissance, 1st edn (1939). The MIT Press (1968), Cambridge Mass

Pedretti C (1999, 2000) Leonardo: the machines. Giunti, Florence

Ramelli A (1588) Livre des Diversse et Artificieuses Machine, Paris (Reprinted by Dover Publishing Company, New York. Gnudi & Ferguson, eds)

Rankine W (1887) A Manual of Machinery and Millwork, 6th edn. Griffin, London

Redtenbacher F (1861) Resultate für den Maschinenbau., Verlag von F. Bassermann, Mannheim, Germany

Redtenbacher F (1862–1865) Der Maschinenbau, 3 Volumes

Redtenbacher F (1866) Die Bewegungs Mechanismen. Verlagsbuchhandlung von F. Bassermann, Mannheim

Reti L (ed) (1980a) The unknown Leonardo. McGraw Hill, New York (see Chapter, The engineer, by Reti L)

Reti L (1980b) The engineer. In: Heydenrich LH, Dibner B, Reti L (eds) Leonardo the inventor. McGraw-Hill, New York

Reti L (1963) Francesco di Giorgio Martini's treatise on engineering and its plagiarists. Technol and Cult 4(3):287–298 (Summer 1963)

Reuleaux F (1875) Theoretische Kinematik; Grundzüge einer Theorie des Maschinenwesens. Verlag Vieweg and Sohn Braunschweig
Reuleaux F (1876) Kinematics of machinery; outlines of a theory of machines (trans: Kennedy ABW). MacMillan and Co, London (Dover Edition 1963)
Reuleaux F (1877) Briefe Aus Philadelphia. Braunschweig
Reuleaux F (ed) (1884) Buch der Erfindungen; Gewerbe und Industrien (The Book of Inventions), reprinted by Bechtermünz Verlag, 1998
Reuleaux F (1885) The influence of the technical sciences upon general culture (trans: from German by Kunhardt W). School of Mines Quarterly, Vol VII, No 1, October (Translation of "Cultur und Technik" in Zeitschrift des Vereins deutscher Ingenieure, 1885, S 24)
Reuleaux F (1893) The constructor; a handbook of machine design, 4th edn. (transl: Suplee HH, 1892), Philadelphia
Reuleaux F (1900) Lehrbuch der Kinematik; Zweiter Band. Die Praktischen Beziehungen Kinematik zu Geometrie und Mechanik, Verlag von F. Vieweg und Sohn, Braunschweig
Reuleaux F (1861) Der Constructeur : Ein Handbuch zum Gebrauch beim Maschinen-Entwerfen, Verlag von Friedrich Vieweg und Sohn, Braunschweig (1st edn. in German)
Scaglia G (1985) Il Vitruvio Magliabechiano di Francesco di Giorgio Martini. Gonnelli, Firenza
Scaglia G (1992) Francesco Di Giorgio: checklist and history of manuscripts and drawings. Lehigh University Press, Bethlehem
Scaglia G (1984) Translation of Taccola, Mariano (1449) De Ingeneis: text and facsimile, Prager UM, Dr. Reichert L, Wiesbaden, Germany
Singer C, Holmyard EJ, Hall AR, Williams TL (1956) A history of technology. Oxford University Press, New York
Sobel D (1995) Longitude. Walker Publishing, New York
Strada J (1617–1618) Künstlicher Abriss allerand Wasser-Wind, Ross-und- Handmühlen, Frankfurt
Strandh S (1989, 1979). The history of the machine. Dorset Press, New York
Taccola M (1449a) De Ingeneis: Text and Facsimile. (See Scaglia (1984))
Taccola M (1449) De Rebus Militaribus: De Machinis. (See Knobloch (1984))
Truesdell C (1968) Essays in the history of mechanics, The Mechanics of Leonardo daVinci, Part I. Springer, New York, pp 1–83
Usher AP (1954, 1988) A history of inventions, revised edition. Dover Publishing, New York (1988)
Valturio R (1492) De Re Militari; Libris XII, 1535th edn. Paris (on line at http://kmoddl.library.cornell.edu; click on references)
Venturi GB (1797) Essai sur les ouvages physico-mathimatique de Leonardo da Vinci avec les fragments tires de ses manuscripts, Vigevano, Guidodi (1335) Textaurus Chez Duprat. Paris
Villard de Honnecourt (1270) The sketchbook of Villard Honnecourt, White T (ed). Indiana University Press, Bloomington (1959)
Pollio V (c.25 BCE) The architecture off Marcus Vitruvius Pollio in ten books (transl: from Latin by Gwilt J), 1826th edn. Priestley and Weale, London, 1874th edn. Lockwood and Co, London (Book X, "On machines and engines").
Von Guericke O (1672) Experimenta Nova, Amsterdam, Holland
Wauer J Mauersberger K, Moon FC (2010) Ferdinand Jacob Redtenbacher [1809–1863]. In: Ceccarelli M (ed) Distinguished figures in mechanism and machine science. Springer, Dordrecht, pp 217–245
Weller AS (1943) Francesco Di Giorgio 1439–1501. The University of Chicago Press, Chicago
Willis R (1841, 1870) Principles of Mechanisms. Longmans, London
Zeising H (1613) Theatrum Machinarum. Leipzig, Germany
Zonca V (1607) Novo teatro di machine et edificii per et secure operationi. Padua, Italy
Zopke H (1896–1897) Professor Franz Reuleaux, a biographical sketch. Cassiers Magazine, New York

Chapter 3
Social Networks of the Internal Combustion Engine and Automobile

Abstract We describe the network of contributors to the creation of the internal combustion engine and its effects on the early automotive industry. This example has links to the aviation network of invention [Chap. 4] as well as to mathematical Chaos Theory [Chap. 6]

Without the internal combustion engine of the late nineteenth century, there would have been no airplanes, no Wright brothers or Henry Ford. It was the enabling technology of aviation and the enhancing technology of the automobile. Steam power required a source of heat, water and a boiler. The higher energy density of petroleum fuels and the higher temperature of the IC engine with its greater efficiency made it the engine choice of aircraft and automobile designers.

The internal combustion, or IC engine, at first can be seen as an evolutionary extension of the steam engine, utilizing the same kinematic mechanisms, materials, manufacturing methods and marketing base. However the IC engine of the nineteenth century embraced the new science of chemistry and combustion. For nearly two thirds of a century, inventors struggled to couple the dynamic chemistry of exploding gases and volatile fluids with the mechanics of moving iron and steel. And although Nicolas Otto [1832–1891] was crowned the genius-inventor who solved the problem in 1884 with his four-cycle engine, there were dozens of attempts over sixty years to create a practical IC engine (Cummins 1989). The solution, using electrically generated sparks to produce ignition, appeared only after the development of electrical and magnetic science and technology.

The steam engine of the eighteenth century was created by craftsmen and artisans, while the internal combustion engine was designed and built by a new breed of polytechnic trained engineers at the end of the nineteenth century. The exceptions were Lenoir and Otto. The evolution of machine design, culminating in the Redenbacher and Reuleaux schools of machine design in Karlsruhe and Berlin led in part to the development of the practical, internal combustion engine. Designers like Benz, Daimler, Maybach and Diesel were trained in technical universities while others like Lenoir, Otto and Bosch came up in the trades.

F. C. Moon, *Social Networks in the History of Innovation and Invention*,
History of Mechanism and Machine Science 22, DOI: 10.1007/978-94-007-7528-2_3,

© Springer Science+Business Media Dordrecht 2014

Many evolving elements of technology led to the rise of the internal combustion engine at the end of the nineteenth century, providing evidence for the avalanche or sand pile theory of invention. In addition to knowledge of dynamics of chemical reactions and electrical science, the following components of the internal combustion engine 'knowledge commons' were essential to its development:

- The improvement of materials such as iron and steel processing for steam engine technology.
- The improvement of machining and casting skills for high tolerance machine parts such as pistons, cylinders and valves.
- The advance of thermodynamic principles such as the law of conservation of energy and the law of entropy and theoretical models such as the Carnot cycle.
- The accumulation of capital in the UK, Europe and United States built on steam, rail and telegraph technologies in the late nineteenth century.
- The development of electrical power technology and the search for portable, non-steam based power units.
- The market need for low cost power units for small manufacturing factories.
- The growing popularity of the bicycle and the improvement of roads.

One could add to this list the rise of aviation and the need for lightweight power units for heavier-than-air machines and dirigibles in the early twentieth century. However as we shall see, the tipping point for the development of the IC engine was around 1880–1890, a decade before the Wright Brothers' flights. At the time of the growth of early aviation from 1906–1914, there were dozens of manufacturers of internal combustion engines in both Europe and North America. The tipping point for the automobile century 1885–1890 followed shortly after the exponential rise of gasoline engine technology.

For the historian searching for data to construct a historical network for IC engines, there is a wealth of published reference materials in English, German and French. Some of this material was published a century ago such as Hiscox (1900), Carpenter and Diederichs (1910) and Clerk (1896). For recent IC history books, in German there is Sass (1962) *History of German Internal Combustion Engine Manufacturing*, and the work of Hartenberg (1999), *The Middle Ages of the Internal Combustion Engine 1784–1886*, translated into English by the Society of Automotive Engineers. From these works we can list the main nodes of an IC network.

Principal Nodes in the Internal Combustion Engine Network:

Jean Joseph Etienne Lenoir [1822–1900]: modified a stationary steam engine topology in 1860 by using coal gas as a fuel; sold several hundred engines.
Nikolas August Otto [1832–1891]: inspired by reports of Lenoir engine, designed his own machine with the help of Langen; developed the first commercial four-cycle engine that became the standard thermodynamic model for internal combustion engines to this day. Franz Reuleaux (Chap. 2) was an early consultant to Otto's efforts.

Fig. 3.1 Photo of Nicolaus Otto [1832–1891] developed the first practical four-cycle internal combustion engine with Eugen Langen. Co-Founded the Deutz motor company in Germany (Sass 1962; Goldbeck 1964)

Eugen Langen [1833–1895]: business partner of Otto; started Deutz engine company; was a friend and fellow student at Karlsruhe with Franz Reuleaux (Chap. 2).
Gottlieb W. Diamler [1834–1900]: worked for Otto and Langen at Deutz; left with Maybach to form a new company devoted to lightweight engines.
Wilhelm Maybach [1846–1929]: worked as an engine designer for Otto and Langen; later went with Daimler to start their own company near Stuttgart.
Robert Bosch [1861–1942]: German mechanic who developed reliable fuel magneto-ignition systems for the internal combustion engine. He spent some time in New York working for the Edison company. He later worked with Otto's Deutz engine company.
George Brayton [1830–1892]: American engineer who invented a liquid fuel internal combustion engine in 1872, called the 'Ready Motor'. The thermodynamic cycle was later used for gas turbines. An oil-burning Brayton engine was displayed at the 1876 Philadelphia Centennial Exposition at which Franz Reuleaux, a consultant to the Otto-Langen company, was the German ambassador.
Karl Benz [1844–1929]: German mechanical engineer who developed an early two cycle gas engine and went on the design and build a gas engine-powered three wheeled carriage. He co-founded the company Benz & Cie. In 1899, they sold over 500 automobiles, and over 3000 vehicles in 1902 a year before Ford started his Model A and Model T company.

Rudolf Diesel [1869–1945]: French born, German educated mechanical engineer who developed a compression-ignition engine, known today as the 'diesel engine'. He studied with Carl von Linde in Munich who had developed refrigeration machines (See e.g. Chap. 7).

3.1 Growth of Technical Events for the Internal Combustion Engine

A measure of the exponential growth of the IC engine is the number of cumulative events such as patents issued. Using data from Carpenter and Diederichs (1910) we see a gradual rise in the early decades of the nineteenth century and a sharp rise beginning in the 1880s decade illustrated in Fig. 3.2.

The exponential growth of the IC engine at the end of the nineteenth century, can be measured in several ways. For example the size of the Otto four-cycle engine in 1878 was 4 BHP, in 1893 it was 200 BHP and by 1908 it was 4000 BHP (Carpenter and Diederichs 1910). This is shown in Fig. 3.3a. When plotted on a logarithmic vertical scale versus years the slope is a straight line as shown in Fig. 3.3b.

As indicated in the opening paragraph, this rise of historical IC events can be correlated with the evolution of both scientific advances in thermodynamics and materials processing as well as manufacturing technologies. Another measure is the growth of US patents issued as listed in Hiscox (1900) shown in Fig. 3.4. Our thesis in this book is that this exponential rise in the measures of innovation is correlated with the growth of a innovation social network.

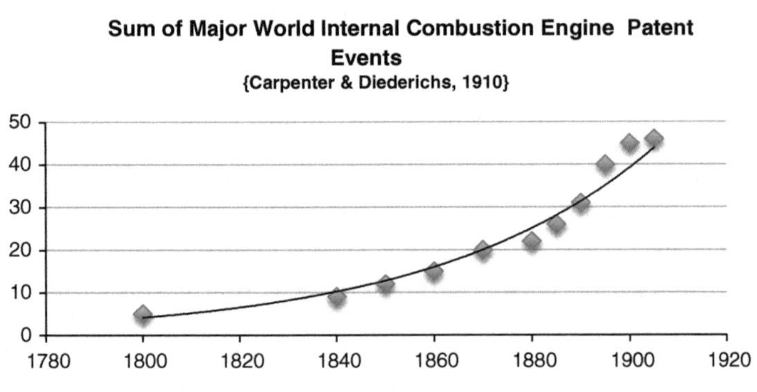

Fig. 3.2 Growth of the number of significant international patents related to the internal combustion engine as reported in Carpenter and Diederichs (1910). The *solid line* is a best fit exponential curve using Excel

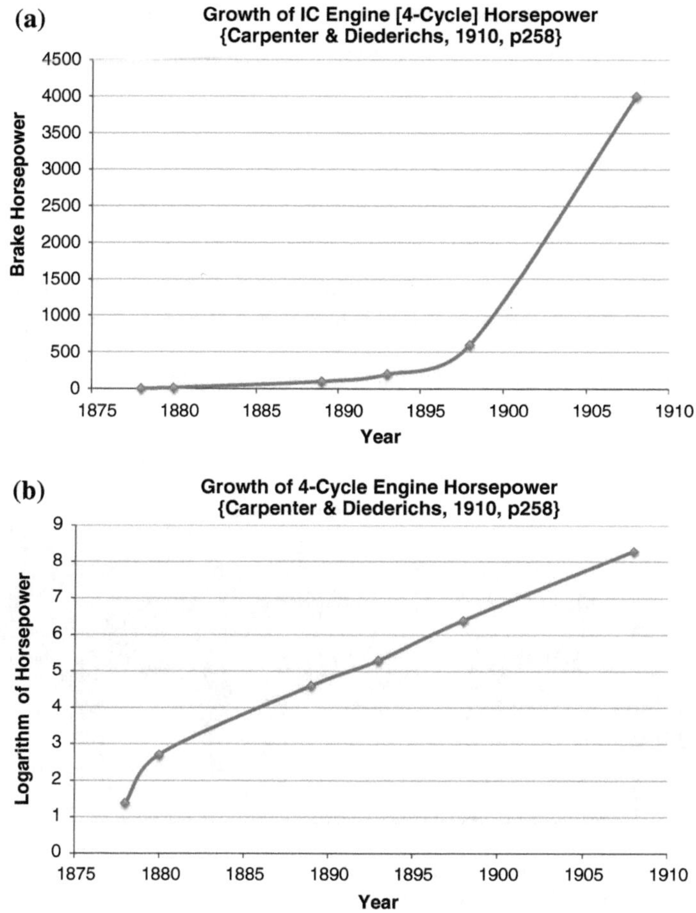

Fig. 3.3 **a** Plot of IC engine brake horsepower (BHP) for Otto engines from 1880–1908. **b** Data in Fig. 3.3a plotted on a logarithmic vertical scale. The *straight line* is an indication of exponential increase in engine power (Carpenter and Diederichs 1910)

3.2 The Standard Model for Internal Combustion Engine Innovation

Before discussing the social network attendant to the internal combustion engine, we summarize the conventional narrative that usually anoints Etienne Lenoir and Nikolas Otto as the genius-inventors of this technology. By the mid nineteenth century, the steam engine had diffused and matured in many applications beyond its original role of pumping water out of the deep Cornish tin mines of England a century earlier. By 1850, the steam engine was a key power source in England, Germany, Belgium and to a lesser extent France. Accompanying this technology, a vast infrastructure

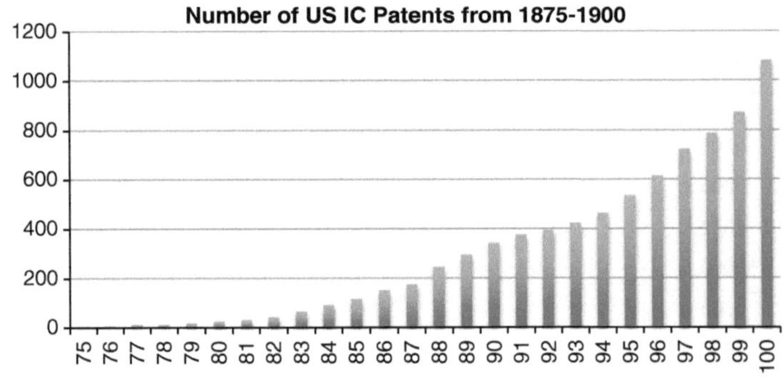

Fig. 3.4 Number of US patents related to internal combustion engines from 1875–1900, showing cumulative patents (Hiscox 1897)

Fig. 3.5 Nicolaus Otto's display of internal combustion engines at the Chicago Columbian Exposition, cover of Scientific American (1893)

of machine shops and machine tool education grew through apprenticeships and polytechnic colleges, especially in England and Germany, out of which came the early successes of Lenoir, Otto, Daimler and Maybach. With the rise of large steam engine powered manufacturing enterprises, there grew a desire to find an alternative power source that was smaller, lightweight and more portable and did not depend on coal and could be used for transportation on land, water and air vehicles. Around 1810, George Cayley, one of the early visionaries of heavier-than-air vehicles, recognized the need for an alternate to steam.

Out of this milieu, Lenoir, using a conventional, horizontal, double-acting steam engine topology, built an engine without a furnace, boiler and steam using an explosive mixture of coal gas. Around 1860, he began to market these small 3 HP engines with great publicity and market interest. As the story goes, Otto read of Lenoir's exploits and sought to make a more efficient design. In 1866, Otto teamed up with Eugen Langen to produce a vertical single-cylinder engine that more than doubled the efficiency of the Lenoir engine and won the Napoleonic prize at the Paris Exposition of 1867. This recognition captured a large part of the market away from Lenoir and established Otto and Langen as major gas engine manufacturers producing over 2000 engines between 1866 and 1876. Otto and Langen exhibited some engines at the Centennial Exposition of 1876 in Philadelphia as well as at the Chicago Columbian Exposition of 1893 (Fig. 3.5). During this time Otto worked to perfect a new engine based on a four-cycle system. Otto and Langen opened up franchises in Philadelphia, Vienna, England and France to become the world's largest manufacturer of gas engines. Otto also hired Daimler and Maybach as engine designers. Daimler became manager of the main manufacturing factory in Deutz, Germany near Cologne.

In the period of Otto patent protection of the four-cycle engine from 1876–1884, other competitors such as Clerk in Scotland began working on two-cycle engines. Also the growing scientific field of thermodynamics brought forth new engines such as the Brayton engine in the United States. As the story continues, Daimler and Maybach split with Otto and Langen and set up a factory in Cannstatt Germany near Stuttgart to manufacture lightweight, high-speed engines for vehicles that ran on petroleum based gasoline. Karl Benz had a similar vision and brought out a tricycle vehicle with gas engine in 1885. Daimler and Maybach showcased a four-wheel vehicle in 1886 and the automobile age was born. By the turn of the century, the gas engine industry was maturing and expanding in a dozen countries with over a thousand patents awarded in gas engine technology in the US alone. This development took place over four decades and was now ready to be exploited by Siemens and his growing electric generator technology as well as by a group of aviation dreamers led by the Wright Brothers.

This narrative gives the primary facts and is compact enough to fit on a page in a book on the history of inventions or machines. But a more detailed reading of primary sources reveals a more complex story. There were many other major and minor players in this evolution that gives a richer understanding of how this key technology evolved and enabled other new technologies dependent on it to also evolve.

3.2.1 A Network Model for Internal Combustion Engine Innovation

Much of the data enabling the construction of the interconnected nodes in the internal combustion engine network comes from early textbooks such as Carpenter and Diederichs (1910) of Cornell University and Hiscox (1900) or the two-volume work of British IC engine pioneer Dugald Clerk (1896). Another valuable source is the scholarly work of Hardenberg (1999) whose detailed description of original patents from the eighteenth and nineteenth centuries has been published by the Society of Automobile Engineers. Also biographies of some of the German players by German speaking authors such as Diesel et al. (1960), provided more links and nodes than

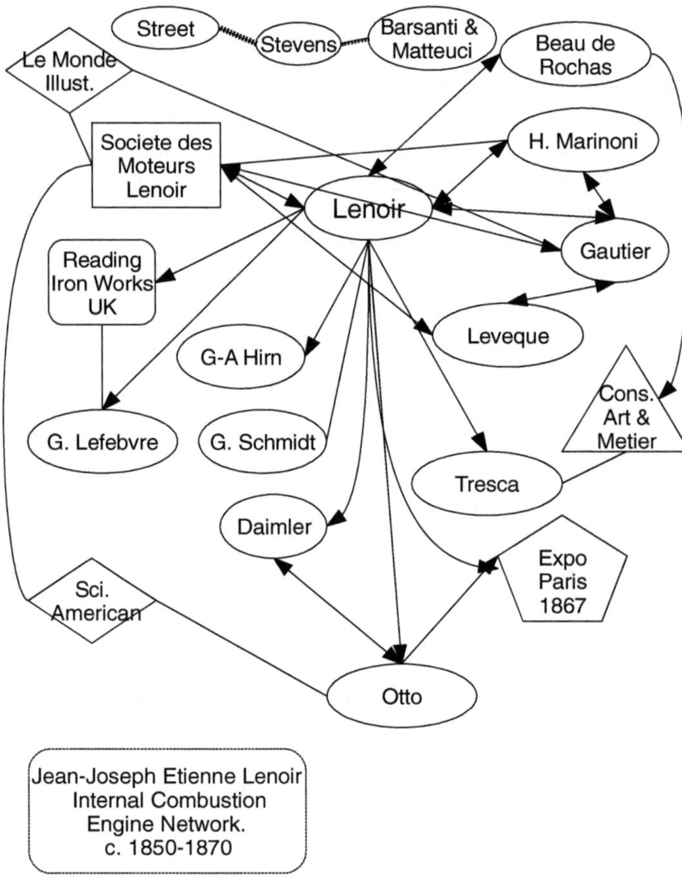

Fig. 3.6 Partial historical network centered around Lenoir his company Societe Lenoir related to his engines of 1860 and 1861 (Cummins 1989; Slade 1866)

that found in thumbnail histories. More difficult to find in English is biographical information about French IC pioneer, Jean Joseph Etienne Lenoir [1822–1900] so the links to his node are not as well known.

A partial network centered around Lenoir and his engine (Fig. 3.6), can be constructed from the Society of Automotive Engineers publication by Hardenberg (1999), pp. 235–266.

An example of a sub-network for the internal combustion engine centered around Otto and Germany is shown in Fig. 3.7, that led to the lightweight engine of Daimler (Fig. 3.8).

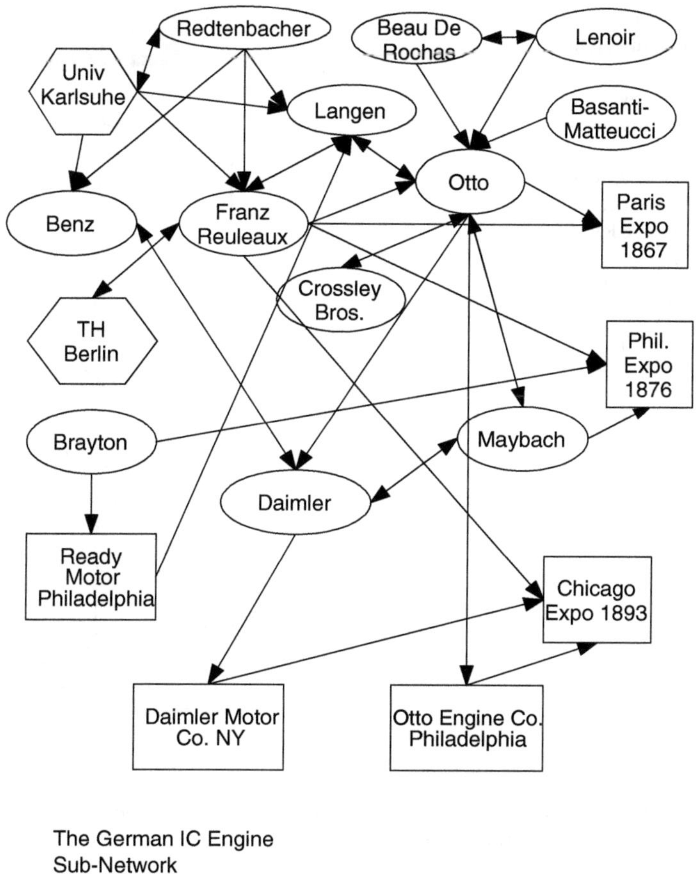

Fig. 3.7 Sub-network for the internal combustion engine 1860–1890, centered around Otto and other German Engineers

3.3 Pre-Lenoir and Otto IC Engines: Linear Growth 1800–1880

The title of Horst Hardenberg's seminal work *The Middle Ages of the Internal Combustion Engine* 1794–1886, gives a teaser to the argument that Lenoir and Otto were not the only nodes in the genius model of IC engine development. Hartenberg begins with a short homage to Leonardo da Vinci and Christiaan Huygens gunpowder engine proposals of the late fifteenth and seventeenth centuries, but then describes and analyzes a dozen or more IC engines spanning six decades where both patent records and working IC models were built. He also uses thermodynamic principles to compare the different machines as to efficiency and power.

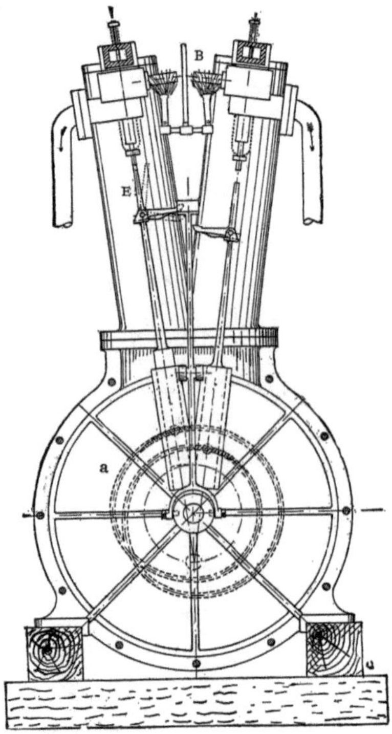

Fig. 3.8 Gottlieb Daimler's two cylinder lightweight internal combustion engine., Cannstatt Germany. (Sibley Journal of Engineering, Cornell University, Ithaca, NY, p. 229, Vol XI, 1896–1897) Specifications; 4.45 HP, 576 RPM, Piston Diameter = 3.425", Stroke = 0.525′. This motor was designed for marine application

These early IC machines include the work of: (See e.g. Carpenter and Diederichs 1910; Clerk 1896; Hardenberg 1999).

Robert Street [1794] British; engine used spirits of tar or turpentine to drive a piston vertically and move a levered pump. The piston was gravity returned.

John Steven [1798] American; Columbia University trained lawyer and designer of steam driven boats built an alcohol fired free piston engine with the help of Marc Isambard Brunel, a British engineer who later achieved much fame in the UK.

Lebon [1804] French civil engineer who patented a gas ignited piston engine.

Niepce brothers [1806] Built a pulverized coal, combustion engine; report given to the institute of France.

William Cecil [1820] British Reverend Cecil built the first hydrogen–air combustion engine and demonstrated it before the Cambridge Philosophical Society.

Samuel Brown [1823], Built and demonstrated, in England and the US, a pistonless gas fired water wheel for canal pumping applications. Patents issued 1823, 1826.

Lemuel Wellman Wright [1834] Coal gas combustion, double acting engine designed similar to a Watt steam engine with water cooled jackets.

William Barnett [1838] Similar to the Wright engine with timed ignition with a rotary slide flame valve. Designed three engines, but no evidence they were built.

Eugenio Barsanti and Felice Matteucci [1856] Spanish teacher Barsanti and engineer Matteucci, built a vertically driven piston engine with a combustible gas–air mixture. Similar to the first Otto-Langen engine.
Pierre-Constant Hugon [1858, 1864] Director of a gas company in Paris patented and built an indirect gas engine where the combustion occurs in a separate chamber. Demonstrated at the 1867 Universal Exposition in Paris.
Christian Reithmann [1860–1869] Bavarian watch maker Reithmann and his technician Ainmiller designed both single acting and double acting gas engines.
George Brayton [1872] American machinist designed and built gas engine more efficient than then popular Lenoir engine. Professor Robert Thurston of Stevens Tech tested the engine and published results.
Dugald Clerk [1880] Scottish chemist designed and built a two-cycle engine.

The names in the above litany of early inventors of IC engines will mean very little to most readers. Greater details about them and their inventions can be found in Hardenberg (1999), available in English translation or German. They are listed here to convey the hidden complexity of the historical innovation networks in Figs. 3.6 and 3.7. They could have been added as nodes to the Lenoir or Otto networks, but discovering the links between these secondary nodes and principal nodes is a difficult task.

A second reason for the mention of these inventors is to emphasize the existence of a *knowledge commons*. The word 'commons' is used here to denote a public space or free information as exists in technical and scientific magazines, or expositions and public exhibitions and demonstrations. A knowledge commons existed in many technologies such as radio electronics or the early days of computer programming. A knowledge commons is also transmitted through 'word of mouth' and is different from the artisan and workshop secret knowledge that was often guarded in the early part of the industrial revolution. Then and now the very filing of a patent disclosed what was possible, such that clever engineers could find ways around the patent. That is why even today some companies prefer the trade secret method rather than the patent disclosure to deter finding paths to their product or process secrets. Although inventors are famous for preferring solitude to human company, they are both curious and clever and will almost always seek information from the knowledge commons.

Following our methodology we plot the number of Hartenberg's internal combustion engine events as listed above versus time, as shown in Fig. 3.9. The graph reveals a linear-like growth in the IC commons knowledge base from 1790–1880.

Although one can approximately fit a linear curve to the data of Fig. 3.9, one might also fit a multi-linear, stepwise curve with plateaus around 1810, 1830, and 1850 suggesting an uneven rate of progress with fits and starts of activity. However this may be taking the data too seriously in that we have a relatively low number of events compared to the later exponential rise of patents shown in Fig. 3.4. The evidence presented by Hardenberg does follow other innovations where there is a linear rate of innovation before the sharp exponential rise that is characteristic of a new revolution in technology. This data also supports the so-called 'sand-pile' model

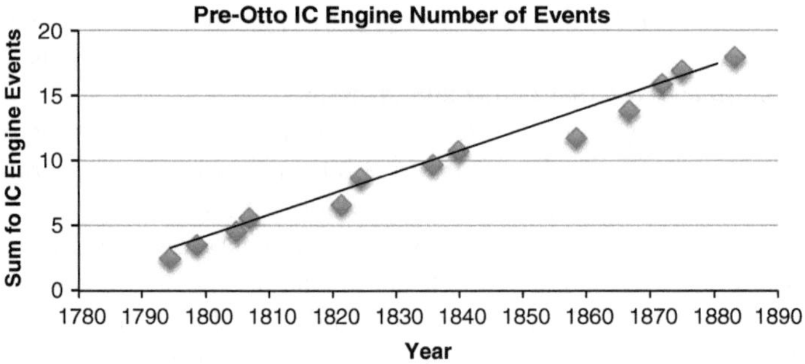

Fig. 3.9 Plot of the sum of major IC events in Hardenberg's *Middle Ages of the IC Engine* (1999) showing a linear or multi-linear progression of the IC knowledge base

of innovation where the knowledge commons accumulates to a critical point where it triggers a sharp growth in the innovation network.

Missing in the IC networks in Figs. 3.6 and 3.7, was the myriad numbers of engine and parts fabricators. Some like Lenoir and Otto were able to secure capital to start their own companies. But other inventors used local machine shops to build prototypes. Also licenses were issued to outside firms such as Francis and William Crossley of Manchester, one of the largest producers of early Otto engines.

3.4 The Evolution of the Automobile

Closely coupled to the rise of internal combustion engine technology is the evolution of the automobile. Again the classic icons in Germany are Benz and Daimler. In the United States, Alexander Winton and the Duryea Brothers were early pioneers but in the history books Henry Ford was anointed as the genius inventor-entrepreneur. As has been shown in the previous examples in this book, the path to the 'horseless carriage' involved many other players, some of whom were part of the IC engine social network.

The automobile was not just an engine on wheels. It was the marriage of a wheeled carriage with a power plant of steam, electric, gas or petroleum. One of the first prototypes was built in 1771 by the French engineer Nicolas Joseph Cugnot [1725–1804], who placed a steam engine on a four-wheeled carriage. Cugnot had studied the machine books of Leupold (See Chap. 2).

The spoked-wheel had undergone more than three millennia of development including the appearance of pneumatic tires in the middle nineteenth century. After the French Revolution, scientific carriage design had also undergone development that produced elliptical spring suspension, lightweight frames, cantered wheels and turntable axles. By the time the IC engine appeared, there was a large industry of carriage makers as well as bicycle manufacturers and artisans with skills to repair both kinds of land vehicles. In addition, the craze for bicycles had created a need for improved roads.

These supporting technologies for the horseless carriage proceeded in a linear progression in the nineteenth century. In 1885, the appearance of Diamler and Maybach's lightweight two-cylinder internal combustion engine, that ran on liquid petroleum, triggered an exponential avalanche in the advance of the motor car as seen in several measures: number of patents, number of manufacturers, innovation event growth or increase of engine horsepower. Thus the social innovation network for the internal combustion engine, that began with Lenoir around 1860, set in motion a closely related social innovation network for the horseless carriage. This rise began around 1885 at least a decade before Henry Ford began tinkering with his friends in his garage with a horseless carriage, and two decades before Ford developed the famous Model T in the United States.

3.5 The Classic Timeline for the Automobile

As we have proceeded in the case of the steam engine and internal combustion engines, we use classic narrative histories as sources to construct timeline growth curves and innovation network diagrams. There are hundreds of books on the automobile as well as sources on the World Wide Web. In this chapter we have used the translation of the German work of Erik Eckermann (2001) published by the Society of Automotive Engineers (SAE), Merki (2002), as well as James Flink's *America Adopts the Automobile* (1970).

Most historical reviews begin with Nicolas J. Cugnot's steam carriage of 1771, and Oliver Evans amphibious vehicle of 1805 (Orukter Amphibolos). Also Richard Trevithick of railroad fame developed a road steam car in 1803. The forerunner of the gas-powered car was Lenoir's three wheeled vehicle of 1863. But steam powered and early gas engine powered vehicles were too heavy and cumbersome. It was the development of lightweight gasoline powered vehicles of Benz, Daimler and Maybach, as well as the development of ignition technology that triggered an avalanche in automotive technology innovation. Although the power plant originated in Germany, the early development of manufactured vehicles began in France, often using German licensed engines. The automobile is an example of what is called today 'systems engineering' and the French had key elements in the automotive system development including better roads than Germany that originated in the Napoleonic era. The French group included Edouard Delamare-Deboutteulle [1884], Panhard et Levassor [1887], Armand Peugeot [1890] who used a Daimler engine, Albert de Dion [1894], Emile Delahaye [1894], Leon Bolee [1896] and the Renault brothers who sold their first car in 1898. The Automobile Club of France was formed in 1895.

In the United States there is the famous automobile patent of George B. Selden [1879–1895] who never built a working vehicle but whose legal maneuvering set back American efforts, a precursor of the Wright Brother's legacy in aviation. The Duryea brothers J. Frank and Charles E. produced a car in 1893 after reading about Benz's vehicle in the 1889 Scientific American. Around 1896 Hiram P. Maxim, Alexander Winton, Charles. B. King and Random E. Olds as well as Henry Ford and his friends build prototype automobiles. Also in 1895 two new automobile magazines appeared, *Motorcycle* and *Horseless Age*, in which they reported over 500 applications to the US Patent office related to automobiles. At this time there appeared a number of electric cars as well as steam cars such as the Stanley Steamer of 1898. By 1899, there were about 30 American automobile companies. In both France and the US automobile racing excited public interest in the automobile, stimulating sales of horseless vehicles.

In 1900 there was an Automobile Show in Madison Square Garden and the Automobile Club of America was formed out of a number of state automobile clubs. In the 1901 Auto Show in New York City there were 58 steam, 23 electric and 58 gasoline cars on show (Flink 1970). But by 1905, there were only 9 steam and 20 electric vehicles compared to 219 gasoline powered automobiles on show. As part of the avalanche of technology in internal combustion engines, engineers could produce lightweight engines of 60 hp in a Mercedes in 1903, 120 times more powerful than Daimler's original 0.5 HP engine of 1885 and well beyond the pace of electric and steam car development.

Using the principal events in the development of the automobile as narrated in Eckermann (2001) and Flink (1970) one can plot the integrated number of events versus time as shown in Fig. 3.10. Like so many other examples of innovation in this book and other sources, the development of the automobile shows an exponential growth in the early period of technical ideas before economic measures show similar growth. In another plot (Fig. 3.11), one can see a growth in cars registered in the US with a sudden rise around 1915 compared with the rise in technical ideas beginning in 1885, a twenty year lag time.

In addition to specific person nodes, industrial company nodes and institutions, such as the Royal Automobile Society, there existed a large knowledge base in internal combustion engine and automotive technology. This knowledge base could be accessed by a novice through magazines and newspapers as well as listening to anecdotal case histories. As prototypes appeared they could not be hidden in a shop; they were paraded on public streets to both public amazement and condemnation. Also parts, tools and automotive supplies could be obtained from a growing number of manufacturers. The automobile is an example of a systems engineering technology that relies not on a singular technical breakthrough, but on the skillful integration of many parts and systems; engine, transmission, electrical system, carburetor, axles and wheels. Much of the initial undercarriage systems were directly adapted from the horse carriage business.

Thus when we read of Henry Ford designing his own four-cycle engine in 1895 or the Wright brothers creating their own lightweight engine in 1903, one must realize that each had access to what we may call a "common knowledge node" in the social innovation network.

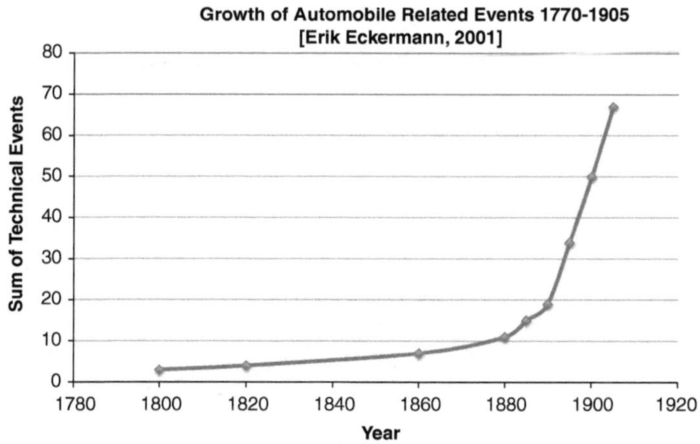

Fig. 3.10 Intergrated number of technical events in the development of the automobile showing a sharp rise around 1885 the time of Daimler and Maybach's lightweight IC engine (Eckermann 2001, Flink 1970, plus several internet websites)

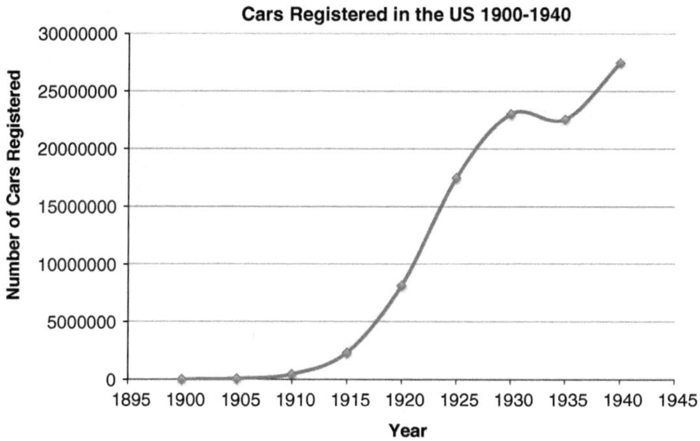

Fig. 3.11 Number of cars registered in the US from 1900–1940 (US Department of Trans. Federal Highway Administration, *The 2000 World Almanac*)

3.5.1 The Henry Ford Model A and Model T Automobile Networks

Using narrative data from a recent biography of Henry Ford by Steven Watts (2006), a partial historical innovation network surrounding Henry Ford is shown in Fig. 3.12. This is necessarily a 'star' shaped network since it is constructed primarily around the life of Ford. A more comprehensive innovation network for automobiles would require the integration with the biographical networks of many other inventors.

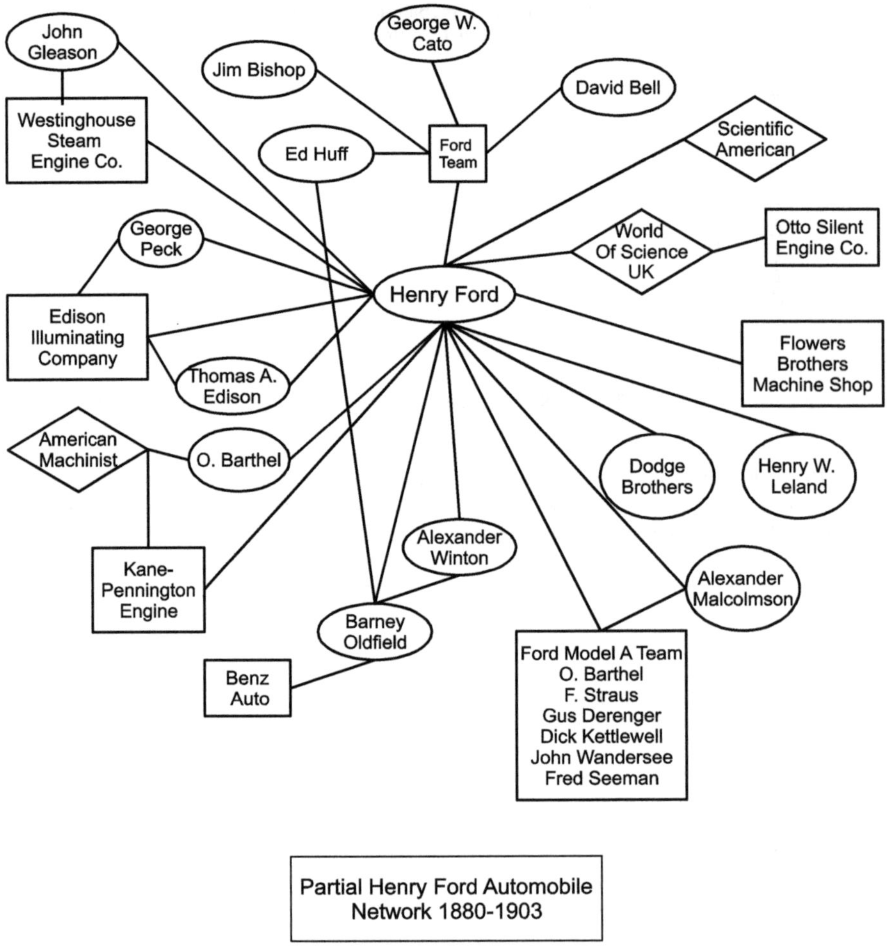

Fig. 3.12 Partial Henry Ford automobile star network based on the biography of Henry Ford by Watts (2006)

A brief guide to some of the nodes in the Henry Ford Network in Fig. 3.12 is as follows:

Otto engine—World of Science and Scientific American; according to Watts (2006), Ford read about the Otto engine in the British magazine *World of Science*. He was also a reader of *Scientific American*, which in its day featured new technical developments such as the lightweight two-cylinder Diamler engine (Fig. 3.8). Also by the mid 1890s, there were new magazines such as *Horseless Age* and *The Automobile and Motor Review* devoted to automobiles.

Westinghouse engine technology; Ford worked for a farmer named Gleason who had purchased a new Westinghouse portable steam engine for threshing and cutting timber. Ford became a demonstrator and mechanic for the machine, taking it around to

different farms. He was subsequently hired by a regional manager for Westinghouse, John Cheeney, to maintain and repair these steam engines. His experience with steam engines on a carriage led him to dream about building a 'horseless carriage'.
Ford backyard garage horseless carriage team; While working for Edison Illuminating Company maintaining steam engines, circa 1896, he brought together a team of mechanics to build a 'horseless carriage' in a shack near the Edison plant in Detroit. The vehicle was called the *Quadricycle,* and was driven around town and attracted a number of investors in the first Ford company that eventually failed.
Edison company sub-network; Through Ford's experience at the Edison Illuminating Company in Detroit, he met Thomas Edison himself and visited Menlo Park. The Edison Company manager George Peck also introduced Ford to future investors in his 'horseless carriage' schemes. Like all successful inventors and entrepreneurs such as James Watt and Samuel Morse, success depended on developing a network of investors.
Automobile racing sub-network; After the failure of his first company, Ford began working on a 80 hp racing car called '999'. He attracted the attention of a then famous race-car driver named Barney Oldfield who had driven for the automobile maker Alexander Winton [1860–1932]. Oldfield had also driven a Benz auto racing car before working for Ford setting a record 131 mph record at Datona Beach Florida. In 1902 the Ford-Oldfield car had a publicized race with one of Winton's cars and won, attracting publicity and new investors for his latest venture the 'Model A'. The following week he incorporated "The Ford Motor Company".
Model A Fordsub-network; The Model A vehicle was built by a team of Ford friends and mechanics colleagues around 1903 with capital from Alexander Y. Malcolmson that launched Ford's automotive empire.

A similar pattern of racing machines and daredevil antics accompanied the early aviation development discussed in Chap. 4. Behind each technology was the push to develop new lightweight, more power intensive internal combustion engines.

According to the integrated event timeline curve in Fig. 3.10, Ford's activities were in the height of the exponential rise in automobile technology. In the early days of the Ford Motor Co. 1903–1906 they produced 13,721 vehicles. However the Old's company produced 17,600 vehicles in this four-year period. From 1906–1910, Ford produced over 83,600 vehicles and Buick would produce close to 60,000 vehicles. Ford would emerge as the leader but about half of American automobiles would still be built by companies other than Ford. While Ford was the 'best among equals' in producing cheap automobiles, he was not the lone genius in the field.

3.5.2 Zipf's law: For Auto Makers

In Chap. 1 we mentioned the research of Yale physicist Derek J. de Solla-Price (1963) who applied statistics to the research publications of scientists. His work followed earlier work of Harvard professor George K. Zipf (1949) who applied statistical ranking to human social entities from word rank in novels to the size of cities. Some of these disparate data sets fall into a mathematical function called *power laws*. When data is plotted on logarithmic scales the points lie on a straight line indicating a mathematical relationship of the form $y = x^{-s}$, where ‘s’ is the negative slope of the Log–Log graph: where ‘y’ is plotted on the vertical axis and ‘x’ is plotted on the horizontal axis.

Recently these power law relationships have been found in the statistics of modern social networks (Bak 1996; Buchanan 2000). In the case of population of cities, this law says that there is no average city. There are a few large cities, a greater number of moderate cities and many more smaller cities. This can be seen for the largest US cities in 1900 shown in Fig. 3.13 top. What is surprising is that this same power law type relation holds for the ranking of US automobile makers in the early years of the industry, shown in Fig. 3.13 bottom. There is no attempt here to claim that the ranking of cities is related to the ranking of auto manufacturers, except that if there is a sociological principle that bigger human groups get bigger and smaller human groups do not get as big, then this may also apply to auto makers. If from 1900 to 2013, New York City has dominated the US population charts in spite of widespread demographic changes, then it is not surprising that Ford dominated the US market from 1906 to 1930. Whatever human dynamics existed a century ago in the decisions as to where people lived, there might have been similar social pressures as to what brands they bought.

What has ranking of automobile brands have to do with historical social networks for the development of the automobile? In Chap. 1 we discussed the construction of social network graphs and how one could assign value or scales to the nodes. In the automobile network, manufacturing companies act as nodes in the same way as person nodes. The ranking of companies by the number of cars sold is one measure of the strength of their node in the network. Similarly one can see that although innovation networks diminish the role of the hero-inventor, the hero-inventor node often dominates the other nodes in the number of links to and from his or her node. In Chap. 4, we show that the number of links versus nodes seems to follow a rough power-law and that the hero-inventor, namely the Wright brothers, had the largest number of links.

The star network of Ford in Fig. 3.12 could be duplicated for a dozen or more principal nodes from Karl Benz and Robert Bosch in Germany to Emile Levassor and Louis Renault in France, and Ransom Olds and Charles and Frank Duryea in the United States. These star networks could be stitched together to form a large quilt with 100 or more nodes and hundreds of links (See e.g. the *World History of the Automobile*, by Erik Eckermann 2001).

We shall not pursue this super network for automobile technology in this chapter. Instead, in the next chapter we will construct a more integrated network for the related technology of heavier-than-air flying machines, employing an influence matrix and applying modern measures of network theory to early aviation. However the path to analyzing the social network for automobile technology has been laid out here.

In this chapter we have tried to make the case for the evolutionary development of the internal combustion engine as a seamless transition from steam engine technology and incorporating thermodynamics and kinematics of machines. From the

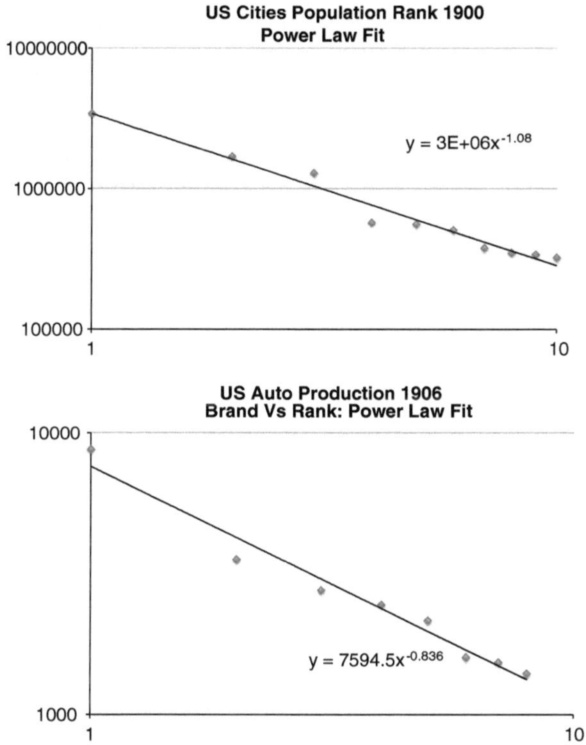

Fig. 3.13 *Top* US population of cities versus rank plotted on logarithmic scales. The top cities are New York, Chicago, Philadelphia, St. Louis. *Bottom* US automobile production for 1906, plotted on logarithmic scales. The brands were; Ford, Cadillac, Rambler, REO Motor Car Co., Maxwell, Oldsmobile, White and Buick

moment a lightweight portable energy source became possible, a worldwide rush to apply this technology began around 1885–1890 with applications to manufacturing, automobiles, electrical power and aviation.

References

Carpenter RC, Diederichs H (1910) Internal combustion engines, 3rd (revised) edn. D. Van Nostrand Co, New York
Clerk D (1896) The gas and oil engine, 6th edn. Wiley, New York
Cummins CL Jr (1989) Internal fire, revised edn. Society of Automotive Engineers Inc, Warrendale
Diesel E, Goldbeck G, Schildberger F (1960) From engines to autos: five pioneers in engine development and their contributions to the automobile industry. Henry Regnery Co., Chicago (German Title; Vom Motor zum Auto: Fünf Männer und ihr Werk 1957. Deutsche Verlags-Anstalt GmbH)
Eckermann E (2001) World history of the automobile. Society of Automotive Engineers, Warrendale
Flink JJ (1970) America adopts the automobile 1895–1910. MIT Press, Cambridge

Goldbeck G (1964) Kraft für die Welt, 1864–1964 Klöckner-Humboldt-Deutz AG. Econ-Verlag, Düsseldorf
Hardenberg HO (1999) The middle ages of the internal combustion engine. Society of Automotive Engineers, Warrendale
Hiscox GD (1900) Gas, gasoline and oil vapor engines, 3rd edn. Wiley, New York
Merki CM (2002) Der holprige Siegezug des Automobils 1895–1930. Böhlau, Wien
Sass F (1962) Geschichte des deutshen Verbrennungsmotorenbaues von 1860–1918. Springer, Berlin
Slade FJ (1866) The Lenoir gas engine, vol 81, 3rd series. J. Franklin Institute, p 178 (Online)
Watts S (2006) The Peoples' Tycoon: Henry Ford and the American Century. Vintage Books, Random House, New York

Chapter 4
Social Networks in Early Aviation History

Abstract We describe a network of over 80 individuals and institutions covering the period of 1810–1910 from the time of George Cayley to the Wright brothers. This case study will provide examples of the importance of 'weak links' and the role of what Gladwell calls 'Connectors' such as Octave Chanute. We also describe in this chapter the use of an influence matrix and statistics of a Link-Node distribution curve.

Did Orville and Wilber Wright invent the airplane? Historians of science and technology have long made a narrative case for the evolution model for scientific and technical innovation in aviation. The aviation historian Anderson (1997) has presented evidence that there were many developments in the science and technology of flying machines before the Wright brothers advances [1899–1904] and it would have been only a matter of one or two years before someone else would have created a controllable flying machine. Our goal in this chapter is to assess the role of social networks in aviation invention and innovation and to seek similarities between the historical aviation network and modern day social networks (Parts of this chapter are taken from Moon 2012).

In this study we have culled a number of classic histories of aviation beginning with Chanute (1896); Gibbs-Smith (1954); Crouch (1981); Anderson (2004), and have created an influence matrix establishing links between approximately eighty nodes spanning the period 1810, the time of George Cayley's writings on flying machines to 1910 the year of the great international aviation meet in Belmont Park in New York City.

For the reader not familiar with the history of aviation, listed below are the major players in the evolution of flight technology;

George Cayley [1773–1857] Early writings on the topology of aircraft.
Otto Lilienthal [1848–1896] Experimented with manned monoplane and biplane gliders.
Samuel Langley [1834–1906] Measured lift and drag and built powered models.

F. C. Moon, *Social Networks in the History of Innovation and Invention*,
History of Mechanism and Machine Science 22, DOI: 10.1007/978-94-007-7528-2_4,
© Springer Science+Business Media Dordrecht 2014

Fig. 4.1 Model of Lilienthal glider in the Long Island Museum of Aviation, New York

Octave Chanute [1832–1910] Published review of the world's research as well as built classic biplane glider.
Gottlieb Daimler [1834–1900] Developed Otto's four-cycle engine into a lightweight engine for flight.
Orville and Wilbur Wright [1871–1948; 1867–1912] Built and flew the first controllable human-carrying aircraft.
Alberto Santos-Dumont [1873–1932] Built flyable aircraft in Europe.
Gabriel Voisin [1880–1973] Designed and built aircraft for many aviators.
Louis Bleriot [1872–1936] First to fly the English Channel.
Glenn Curtiss [1878–1930] Developed lightweight engines and fast airplanes as well as built the first seaplanes.

These principals were linked by a historical network of more than seventy other individuals and institutions.

4.1 A Social Network of Early Aviation

'*Historical innovation networks*' consist of a set of persons (inventors, scientists, artisans), organizations (e.g. universities, Royal Societies) and social events such as international expositions, including aviation racing tournaments. Unlike modern scientific networks with citations and email, aviation network links between person-nodes were created by friendship, letter exchange, writings, books, and financial interests, to mention a few. Connections between event nodes and person-nodes were created by attendance or by proceedings etc.

Links in the aviation network are either directional or bidirectional. For example links between nodes in different generations are unidirectional. Lilienthal could influence the Wright brothers, but they did not influence Lilienthal because he died in 1896 3 years before the brothers became interested in flight in 1899. In social net-

Table 4.1 Sub-matrix of influence links; rows are the source nodes (outward links) and columns are the dependent nodes. The "link" column sums the number of influence links for a given source node

Node	Cayley	Lilienthal	Chanute	Wrights	Bleriot	#Links
Cayley	X	1	1	0	0	2
Lilienthal	0	X	1	1	1	3
Chanute	0	1	X	1	1	3
Wrights	0	0	1	X	1	2
Bleriot	0	0	0	1	X	1

work theory, nodes can have attributes called *fitness*,which measures the importance of the node or link. For example the Wright brothers would be a high fitness node as would be Lilienthal. Unlike social networks such as the World-Wide Web, the aviation network spans generations linking person-nodes living in one generation with dead person-nodes in an earlier generation. The aviation historical network evolved as new nodes were added and new links are created between new players exhibiting exponential growth in the period 1895–1910. There were also *sub-networks* focused on specific components such as lightweight engine evolution, fuselage structure or wing design as well as the underlying aerodynamic theory.

An overall picture of early aviation network is shown in Fig. 4.2. For print legibility, only about 60 % of the nodes are shown and about 50 % of the links. Even at this level one can see a significant complexity in this network. Not shown here is the directedness of the links or the fitness or weighted value of the nodes and links. Lists of person nodes, association nodes and university nodes are given in the Appendices for Chap. 4.

The data in Fig. 4.2 was obtained from an influence matrix M(NS, NT) where NS are the source nodes of the influence and NT are the terminus nodes of the directed influence links. Thus Cayley is an NS link for Chanute, while Chanute is a source node for the Wright brothers. The matrix M is 80 $\times$ 80 nodes and the entries are not weighted and contain only zeroes and ones. (See Table 4.1.) In general node-pairs that were contemporary, such as Lilienthal and Chanute in Table 4.1, were assigned both outgoing and incoming links; it was assumed that each influenced the other. Earlier generation nodes connecting later nodes, such as Chanute and Bleriot in Table 4.1, were assumed to have only outgoing links. A more detailed historical network would include weighted, as well as directed links. One could also scale the importance of the nodes with respect to the advance of the technology, however we did not attempt to assign a node scale in this study. An example of a sub-influence matrix is shown in Table 4.1.

The top of the network in Fig. 4.2 dates from 1790 and the Montgolfier brothers node and 1810 with George Cayley node. The lower middle section around 1890 has Otto Lilienthal and Octave Chanute as major nodes and the lower part of the graph dating from 1900 to 1910 has nodes for Orville and Wilbur Wright in the US and Gabriel Voisin and Louis Bleriot from France. The reader can find a list of short biographies of most of these nodes in Longyard (1994). Also shown in this graph are a few institutional nodes such as the Aeronautical Society of Great Britain (later the Royal Aeronautical Society) founded in 1866, the Technische Hochschule Berlin, where Lilienthal attended, Cornell University in upstate New York State where a number of secondary but essential nodes were associated, and the Smithsonian Institution, where Langley was Director and from whom the Wright brothers obtained their early information. (See e.g. Mackersey 2003.)

This network includes only a few ballooning nodes. A 'lighter-than-air' machine network described later in this Chapter shows complexity similar to that for 'heavier-than-air' machine development. However a number of ballooning nodes are included in the network of Fig. 4.2. Cayley was inspired by the ballooning exploits of the Montgolfier brothers. Also, the American balloonist T. Lowe mentored Zeppelin. Zeppelin later employed a Daimler engine in his dirigible that led to lightweight engines for 'heavier-than-air' machines. Similarly, the American balloonist, Thomas Baldwin used a Curtiss motorcycle engine in his dirigible that brought Curtiss to the attention of Alexander Bell. In 1907, Bell assembled American and Canadian aviation enthusiasts into the *Aerial Experiment Association* at Curtiss' workshop near Hammondsport NY, to challenge the Wrights' claim of manned flight.

Another measure of the progress of heavier-than-air development is a timeline plot (Fig. 4.3) and the growth of event nodes with time, shown in Fig. 4.4a, b. The addition of new nodes shows the beginnings of exponential growth several years before the seminal year of 1903 in which the Wright brothers demonstrated a controllable flying machine.

In a recent lecture by the author, one commentator suggested that the dual slopes in Fig. 4.4b, might be interpreted such that the low slope growth reflects the "passion years" of aviation and the steeper slope the emergence of the "profit years" through air meet prizes and aircraft sales.

As historical networks advance, not only does the number of nodes grow, but so also does the number of links between the nodes. And in the age of the telegraph and the early telephone age, this growth in both the aviation nodes and links occurred in an exponential way as illustrated in the Fig. 4.5. In this network the number of nodes grew from the time of George Cayley [c. 1810] to the time of the Wright Brothers, including over 40 nodes and generating up to 220 one-way links between the nodes.

We can also associate with many of the nodes the creation of actual flying machines. A plot of the growth of engine-powered flying machines in the period 1900–1915 is shown in Fig. 4.6 using data from Jane (1917) *All the World's Flying Machines.* This illustrates the fact that, although the Wrights did not fly publicly until 1908 in Paris, the world's knowledge base in aviation science and technology was ripe by the turn of the century, enabling the creation of dozens of aircraft in a half dozen nations.

The exponential rising curves in Figs. 4.4 and 4.5 consistently show that by 1900, the avalanche of events, players and links was well underway before the Wrights' famous flights of 1903. If a wind storm had destroyed their aircraft on the dunes of Kitty Hawk that December, within a year or two someone else would have flown a controllable aircraft. The exponentially rising graph of Fig. 4.6, showing the increase

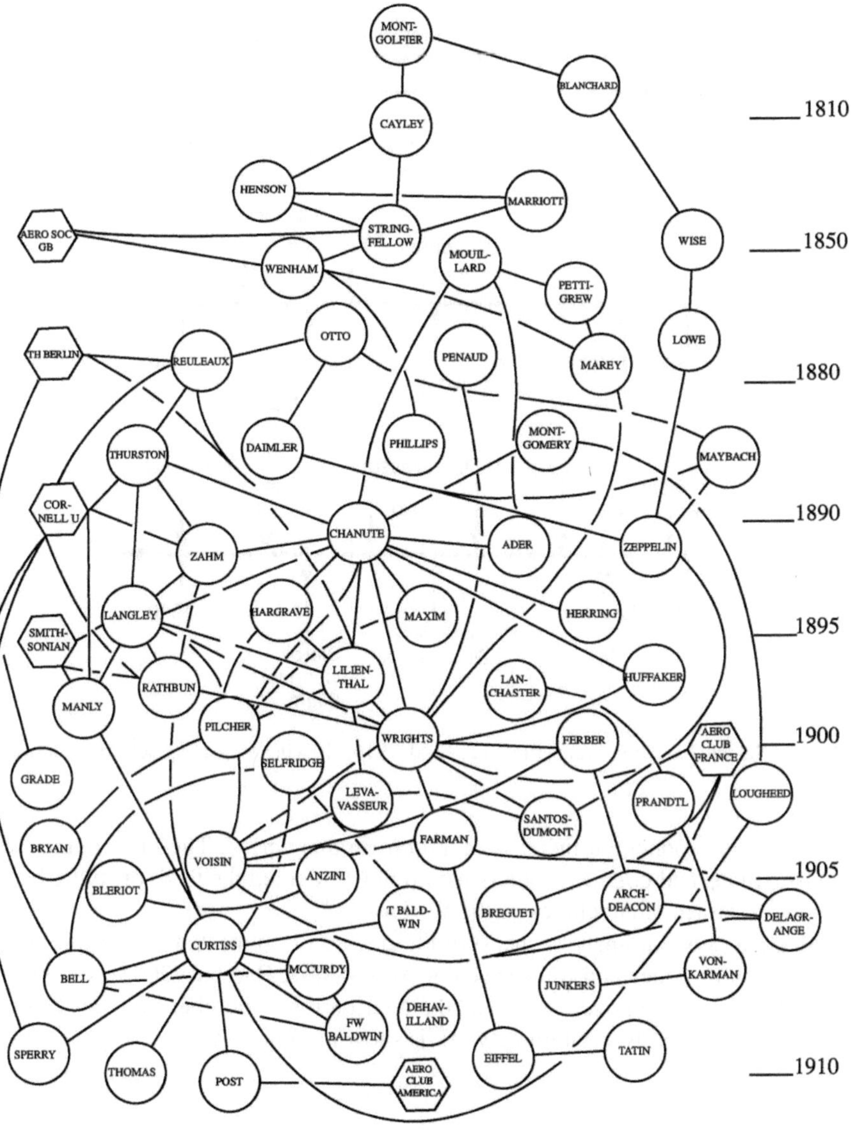

Fig. 4.2 Partial aviation network of innovator nodes and selected links, including six institutional nodes, for the period 1810–1910. Directional nature of the links is not shown. Most ballooning innovation nodes are not included

in flyable aircraft between 1903–1910, certainly was influenced by the Wrights' remarkable accomplishments even though there were no public demonstrations by the Wrights until 1908. Just the news that they had done it was enough to encourage others to risk capital and life. Also patent threats from the Wrights seemed to have not

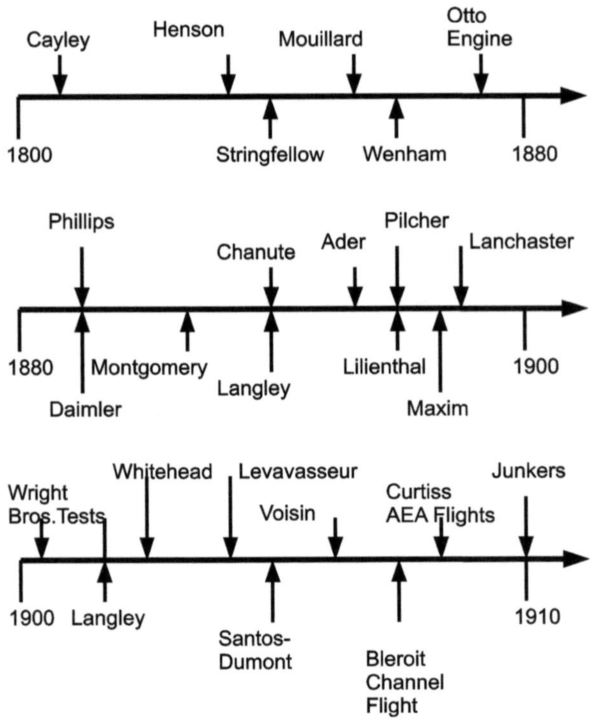

Fig. 4.3 Selected aviation events timeline, 1800–1910, showing increasing density with time

impeded the exponential increase in aircraft in this period. The French, the British the Brazilians and countless others would not be denied the thrill of flying because of some abstract patent claims of the Wrights.

Modern network analysts measure the degree of connectedness by plotting the distribution of nodes and links (See e.g. Watts 2003; Barabasi 2003). Using a data base of 66 nodes and over 300 links, the distribution of number of nodes N(k) with 'k' links for our early aviation network is shown in Fig. 4.7.

4.2 Discussion of Results

The nature of modern networks such as airports in the US or the World-Wide Web, has been characterized as 'scale-free ' or power law networks (See e.g. Watts 2003; Barabasi 2003) in which a small number of nodes have the majority of links and the majority of nodes have a small number of links. This behavior seems to be similar to the early aviation network of Fig. 4.2. (In social network theory however, the negative exponent of the number of links 'k' in the power law is greater than two whereas the exponent in Fig. 4.7 is one. This might be a consequence of the low number of

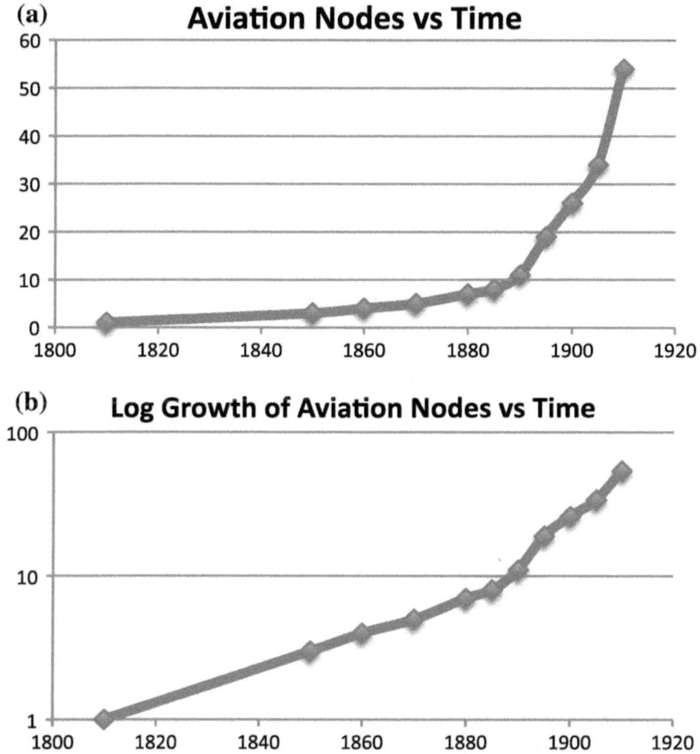

Fig. 4.4 **a** The growth of the number of aviation innovation nodes versus time. **b** Bottom Logarithm of aviation nodes growth with two *straight lines* showing periods of different exponential growth

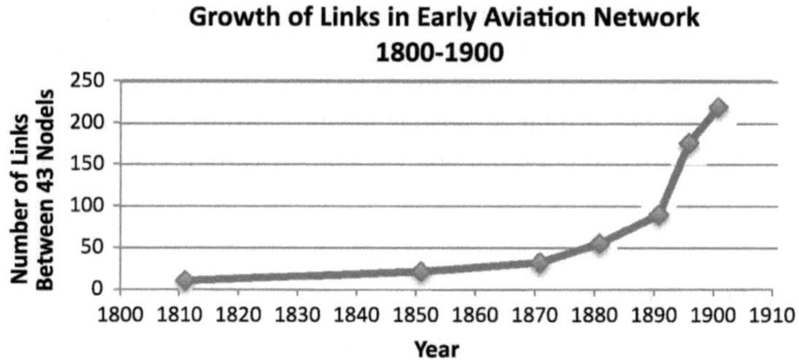

Fig. 4.5 The growth of one-way links in early aviation technology as the number of innovator nodes grew from one node (George Cayley) to the Wright Brothers for a total of 43 person nodes

aviation nodes relative to large networks such as the World-Wide Web. See Newman (2003) for a more detail discussion of power laws in social networks.)

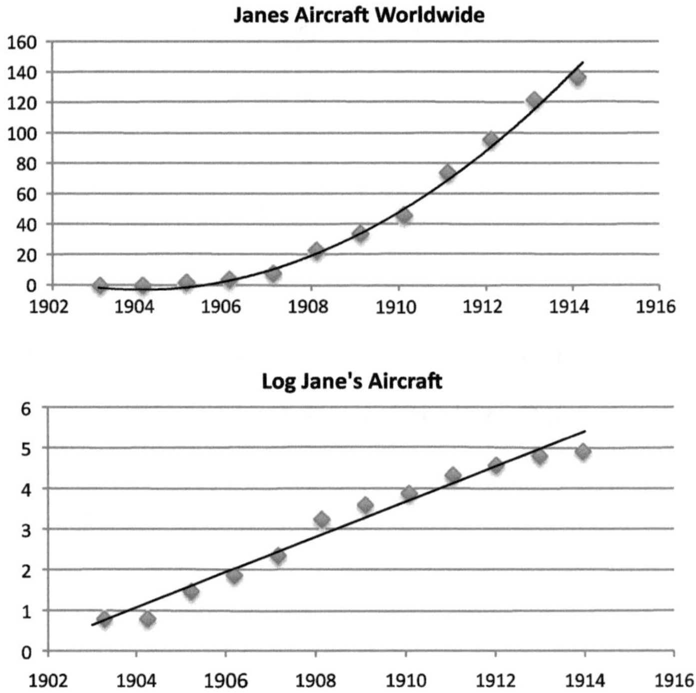

Fig. 4.6 *Top* growth of flyable human carrying aircraft in the post Wright brothers era. From Jane (1917). *Bottom* Logarithm of vertical axis of data in Fig. 4.6 top. *Straight line* represents an exponential increase in aircraft in the decade after the Wright brothers first flights

The two largest nodes are the Wright brothers and Octave Chanute with 24 and 23 links respectively. The Wrights influence is understandable. However, Chanute's role as one of the most important nodes in the aviation social network is largely ignored in popular histories except that of serious aviation historians. (At this writing the author could find only one biography on Chanute by Simine Short (2011)).

4.2.1 Octave Chanute

Octave Chanute (1832–1910) was born in France and brought to the United States as a boy by his father (Short 2011). In New York State he was apprenticed to railroad engineers and eventually was responsible for extending rail systems all over the Midwest. He retired in his sixties and pursued his passion for flying machines. To illustrate the activity in aeronautical technology and science a dozen years before Orville and Wilbur, in 1889 there was an *Exposition Universelle de Paris* at which was held a *Congres International d' Aeronautique.* Over a hundred participants attended including Chanute and Langley. The first speaker was Ettienne Marey who

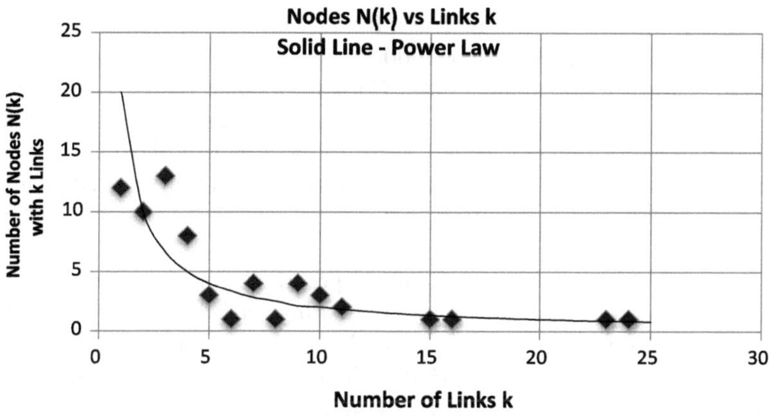

Fig. 4.7 Distribution of number of nodes N(k) with 'k' unidirectional links for the aviation network in Fig. 4.2. *Solid Line*; Power law, N(k) = 20/k, for the data in Fig. 4.7. Number of nodes = 66 and number of links = 320

had studied the flight of birds. Marey's book *Animal Mechanism* was in the library of the Wrights as young boys (Crouch 1989). Little did Chanute imagine that in 1893 he would host a similar symposium on aeronautics in Chicago.

The importance of international meetings in creating innovation networks is illustrated in many of the case studies in this book. Another venue is professional societies. In Chanute's case it was the American Society of Civil Engineers where he met Robert Thurston. Chanute became the President of ASCE in 1891 and Thurston was president of the American Society of Mechanical Engineers in 1882. ASME would become the professional home to some of the aeronautical engineers in the early decades of the twentieth century. Chanute and Thurston were also members of the American Association for the Advancement of Science or AAAS. In 1886 at an AAAS meeting in Buffalo, Chanute had invited a speaker to talk about flying models and birds. The talk was controversial but in the audience was Samuel Langley who became intrigued by the lack of hard science in the principles of flight.

After his trip to Paris, Chanute communicated with almost all the principals in early aviation and wrote a series of articles tracing the history and progress in the *Railroad and Engineering Journal*, published by his friend Matthius Forney whom he knew from ACSE. In 1890 Thurston invited Chanute to lecture at Cornell Sibley College of Mechanical Engineering on progress in aviation. (See article in the Sibley J. of Engineering, 1890.) These lectures and articles were repackaged in Chanute's groundbreaking book, *Progress in Flying Machines,* in 1894. This book was also read by the Wright brothers, c. 1899, before they built their first gliders.

Chanute began to build and test airplane models. At that time there was no agreement on the topology of a heavier than air machine. Lilienthal had wings shaped like birds (Fig. 4.1). Langley experimented with flat and curved planes. Chanute constructed models with eight, four and two cambered planes, before settling on a biplane with cambered wings. The two planes were separated with struts and wires in what civil engineers call a Pratt Truss. It was with this topology that the Wright brothers began their own research in 1900.

In 1896, Octave Chanute along with Cornell graduate Alfred Zahn, organized an *International Conference on Aerial Navigation* similar to the Paris meeting, at the Columbian Exposition in Chicago. Many of the principal players in aeronautics were present. Thurston attended as did Alexander Graham Bell. The German, Franz Reuleaux, attended the Chicago Expo but it is not known if he attended Chanute's symposium.

Perhaps inspired by the news of Lilienthal's progress with gliders, in 1896 Chanute decided to conduct full scale experiments on soaring and hired young assistants such as Augustus M. Herring to make over 700 flights on the dunes of Lake Michigan. The links here are also interesting. Herring had studied at Stevens Institute around 1893 where Thurston had taught before coming to Cornell in 1885. Herring was inspired by the work of Lilienthal to take up gliding. Herring had also worked for Langley. In 1895, Langley had gone to Berlin to see Lilienthal fly from his man-made soaring hill. As mentioned in Chap. 2, Thurston had met Franz Reuleaux in Berlin in 1878, and Reuleaux was a mentor to Lilienthal some years later. So the seeds of American-German engineering connections were sown 15 years before Langley and Chanute become interested in German aviation. As mentioned above, Thurston was also friends with Langley. Later Herring would partner with Curtiss in his aircraft manufacturing venture but that is too complicated a story for this Chapter (Shulman, 2002).

One cannot recreate all the specific details of the links in this sub-network of American aviation but the network was beginning to grow rapidly by the mid 1890's. Of course similar complex direct and indirect links in the aviation network in France and the UK were also underway. Clemont Ader in France had built and tested a strange looking machine called *Eole* in 1890 and Hiram Maxim who had lived in Brooklyn running a manufacturing firm, went to England to built his own aircraft in 1894. Chanute reported on both of these flight attempts as well as others in his articles. Less than a decade later, after Chanute had published his book on flight, he exchanged hundreds of letters with the Wright brothers on the technical details of designing and building flying machines (MacFarland 1953). In the Alfred Zahm Archives at Notre Dame there is a letter from Chanute to Zahm, dated Jan 9, 1904, less than a month after the Wrights' famous Kitty Hawk flights.

"My dear Mr Zahm,

I enclose a statement by Wright Brothers of their real performances with a flying machine, to which the newspapers do not seem to give as much publicity as to the former sensational reports.—"

Along with the letter is a Chicago Daily News article dated January 6, 1904. "*Wright Brothers Give Account of Airship Test in North Carolina*", with a by-line from the Associated Press. The article runs about 20 column inches and is quite detailed. However it ends with a quote from the brothers that is a prelude to many of the problems they ended up creating for themselves.

"From the beginning we have employed entirely new principles of control and as all the experiments have been conducted at our own expense, without assistance

Fig. 4.8 Octave Chanute [1832–1910] Chanute and the Wright brothers exchanged hundreds of letters between 1900–1910. (MacFarland 1953; Crouch 1989)

from any individual or institution, we do not feel really at present to give out any pictures or detailed description of the machine."

After the Wrights had access to Lilienthal's flying data, Langley's wind tunnel testing at the Smithsonian Institution and Chanute's biplane Pratt truss design, to disclaim any assistance from "any individual or institution" took a great deal of denial. Inventors rarely see themselves as part of a greater network seeking the same goals.

A sub-network centered around Octave Chanute is shown in Fig. 4.9. Here we include more detail but we omit links to other nodes outside Chanute's influence. As evidence of the connectedness of this network we quote a letter from Robert Thurston of Cornell to his former student Alfred Zahm, at Notre Dame, dated May 16, 1892 in response to Zahm's invitation to attend the Chicago Conference of Chanute.

"*Your (letter) of the 14th is received. In reply I would say:*

That the institution of an Aeronautical Congress would seem to me to be likely to prove an admirable move. I should think it certain to attract a great deal of attention and a large attendance.

I would suggest the invitation of all members of the various aeronautical societies, of men like Langley, Maxim, Hargreaves (sic) and others who have actually done something to advance knowledge in that field, and to make the proposed invitation widely known through publications, like the Scientific American, which goes all over the world. I shall certainly attend, if possible and will endeavor to prepare a paper.

Yours truly, R.N. Thurston"

The Chanute sub-network in Fig. 4.9 shows much greater detail than Fig. 4.2. The graph has a more 'star'-like shape. When this graph is sewn together with other sub-networks, such as the Wrights' or the French aviators', the overall network has the complex qualities illustrated in Fig. 4.2.

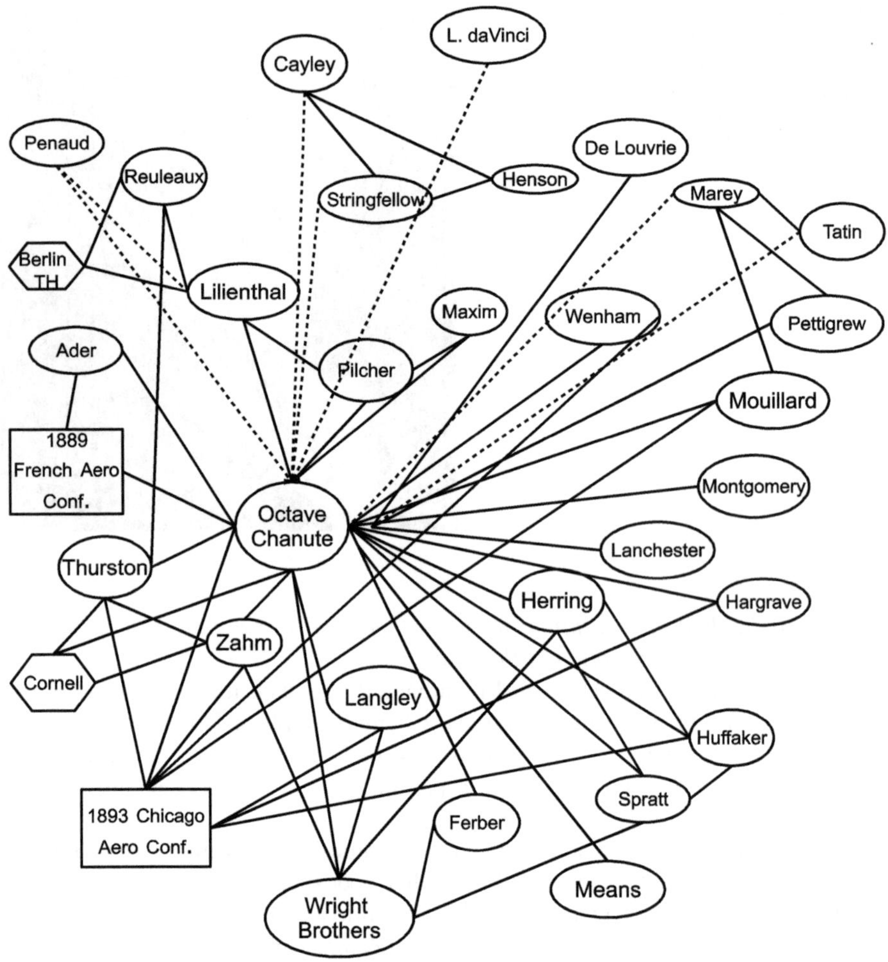

Fig. 4.9 The Octave Chanute sub-network spanning 1800–1905

The two next largest nodes in the aviation network in Fig. 4.2 are Otto Lilienthal 1889 (who died in 1896 in a glider test) and Samuel Langley (1911) with 16 and 15 influence links respectively. Both innovators had published articles and books on flying machines that served to extend their influence. Although Langley's 1903 machine did not fly, he had performed careful experiments in the early 1890's and had succeeded in creating large unmanned engine-powered flying machines in the late 1890's that gave encouragement to the growing number of inventors and would-be aviators.

Further down the list in Appendix I are Lawrence Hargrave and Robert Thurston with 10 influence links. Although Hargrave worked in Australia, he belonged to a local scientific society and published more than a dozen papers on flying which he sent to others in Europe and North America including Robert Thurston of Cornell.

The Robert Thurston node is an example of an enhancing node—one that is not a principal contributor (he presented a paper on "materials of aeronautic engineering" at Chanute's 1896 Chicago Conference) but one that linked important players together. Thurston was an expert on the steam engine and thermodynamics and taught at the US Naval Academy and Stevens Tech before coming to Cornell in 1885. Thurston invited Octave Chanute to Cornell in 1890 to lecture on aerial navigation when Chanute was worried that he would be ridiculed for speculation about flying machines. Alfred Zahm (2003) was Thurston's student who attended these lectures, and graduated from Cornell in 1892. Zahm convinced Chanute to organize an international conference on aerial navigation in 1893 at the Columbian Exposition in Chicago. Around 1900, Thurston was asked by his friend Langley to recommend a Cornell engineering graduate for the scale-up project of Langley's flying models and Charles Manly took the job.

Thurston was also a friend of Alexander Bell who was also a Cornell lecturer. Bell later organized the Aerial Experiment Association in 1907 out of which came Glenn Curtiss's 'June Bug' machine. (See e.g. Curtiss and Post 1912; Hatch 1942.) Thurston also had links to the Berlin nodes of Reuleaux and the engine nodes in Germany. William Durand was a younger colleague of Thurston at Cornell who would later go to Stanford, start propeller research and head up the new National Advisory Committee on Aeronautics in 1914. Another Sibley College student from Cornell was Elmer Sperry, who with his son developed a gyro controller for aircraft. Lionel Marks another Cornell grad taught at MIT and wrote an early textbook on aircraft engines.

The Cornell University aviation node shows the increasing importance of research universities in technology innovation networks beginning at the middle of the nineteenth century. In Chap. 2 we mentioned the importance of the German polytechnic colleges in the creation of the internal combustion engine. In Chap. 5 Columbia University in New York City is shown to be an important node in the radio electronics network. In the twentieth century, Stanford University was an important node in robotics technology.

Another feature of the early aviation network are so-called 'small-worlds' or subnetworks defined in the paper of Watts and Strogatz (1998). Examples are the troika of Stringfellow-Henson-Marriott who were influenced in the mid nineteenth century by the writings of Cayley (1810). Chanute in his classic work, *Progress in Flying Machines, 1894,* provides a link to the work of this troika and other late nineteenth century aviation innovators. Another small world group is the Alexander G. Bell Aeronautical Research Association that enlisted Glenn Curtiss. The French team of Voisin-Bleriot-Levavasseur were linked to a wider network by Archdeacon, the President of the French Aero Club. Bleriot was the first to fly the English Channel in 1908, in a airplane built by Voisin and an engine by Levavasseur.

One of the interesting nodes is American James Means [1853–1920] who published the *Aeronautical Journal* in the years 1895, 1896, 1897. After a year at MIT, Means became a self-made Massachusetts businessman who made his money manufacturing shoes. (Means 1964) Means retired early to pursue his interest in flying machines. He became a friend of Chanute and through his influence, publicized the work of Lilienthal, Langley, Wenham, and Maxim in his journal. He republished the early work of Cayley. He and a few Boston friends formed the Boston Aeronautical Society in 1895. He believed in the value of building flying models and especially the work of Langley. During the summers Means built dozens of gliders of the order of 4 ft in scale and crashed many in trying to replicate the secrets that Lilienthal and others had recently discovered. The Aeronautical Journal was very influential and was read by many including the Wright brothers as they later acknowledged a decade after their successful flights. Means' Journal as well as others in Europe spread the word of the excitement of building flying machines, educating the public into the past history and current research.

Associations and Print Journals

There were institutional nodes such as the Aeronautical Society of Great Britain founded in 1866 or the Aero Club de France of importance to the aviation network. Many of the German Universities had glider associations. In the United States several Ivy League Colleges had glider 'aero' clubs before WWI, including Cornell University. (NY Times, June 1, 1913) See Appendix II for a list of early aeronautical societies, as well as Fig. 4.11. Some of these aeronautical associations were focused on ballooning as well as heavier-than-air flight. There were hundreds of members in these aero clubs who helped spread the word that human flight was not only possible but was close to reality at the turn of the twentieth century. (See Appendix II, this chapter.)

Other nodes, not shown but equally important, were journals such as *Scientific American*, the *American Engineer and Railroad Journal*, and *Means Aeronautical Journal* to mention a few in the United States and *L'Aerophile* in France. Another transient node class was international meetings, such as one held in France in 1889 which Chanute attended and the 1893 Third International Conference on Aerial Navigation in Chicago that Zahm and Chanute organized. Chanute's linguistic ability in French and English also helped spread the news of US progress to Europe as well as interpret research from such obscure nodes as Louis-Pierre Mouillard in French-controlled Cairo, Egypt.

Sibling Pairs and Birth Dates

An odd collection of important nodes are those with Sibling pairs. The Wright brothers pair is the most famous but there were others, such as the Lilienthal brothers, Farman, Voisin and Lougheed brothers and predating all these the Montgolfier brothers. In addition there were several brother and sister nodes and sisters nodes such as the 'Flying Sisters' Marjorie and Katherine Stinson (See e.g. Longyard, 1994) Sister Ella was a close assistant to Percy Pilcher before he perished in a glider experiment. A list of 'Sibling Nodes' is given in Appendix IV for Chap. 4.

An interesting demographic measure is the distribution of birth dates of human nodes. The double peaked distribution in Fig. 4.12 has a group born around 1840 such as Chanute, Langley and Maxim, and another peak a generation later around

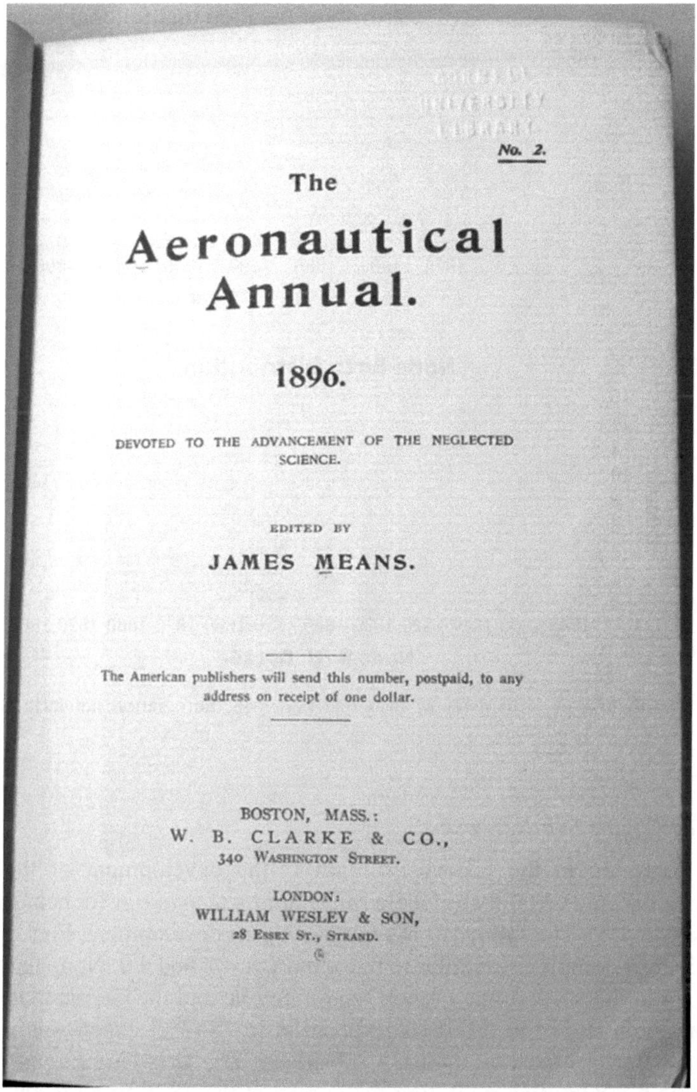
No. 2.

The

Aeronautical
Annual.

1896.

DEVOTED TO THE ADVANCEMENT OF THE NEGLECTED SCIENCE.

EDITED BY

JAMES MEANS.

The American publishers will send this number, postpaid, to any address on receipt of one dollar.

BOSTON, MASS.:
W. B. CLARKE & CO.,
340 WASHINGTON STREET.

LONDON:
WILLIAM WESLEY & SON,
28 ESSEX ST., STRAND.

Fig. 4.10 Facing page of James Means *Aeronautical Annual, 1896.* In this issue Means presents articles on the flights of Lilienthal and the experiments of Hiram Maxim. Maxim who built a lightweight steam engine to power his machine, recommended that younger aeronauts consider gasoline fueled power plants. (See Means, 1964 for a list of articles for all three issues)

1870–80, including the Wright Brothers, Curtiss, Voisin, Bleriot. The first group might be called the "dreamers" and the later group the doers.

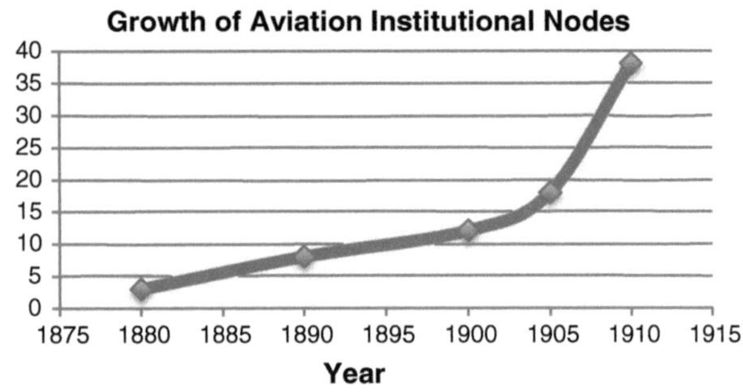

Fig. 4.11 Growth of Aviation Related Institutional Nodes (See e.g. Moebedeck 1906)

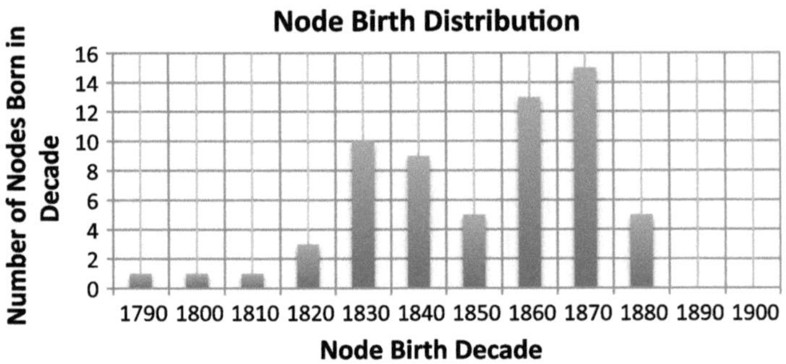

Fig. 4.12 Distribution of birth dates of principal nodes in the aviation network; See Chap. 4 Appendix

The Gas Engine Sub-Network

Often forgotten in the history of flight is the development of the gasoline engine, which would certainly be labeled an *enabling technology* for heavier-than-air machines (Chap. 3). The sub-networks for gas-engine development, Figs. 3.6 and 3.7 in Chap. 3 show complexity similar to those in Figs. 4.2 and 4.9. Non-steam engines have origins in the work of the French Lenoir engine and the German Otto-Langen gas engine, both shown at the Paris Exposition in 1867. (Langen was a friend of Reuleaux in Berlin.) Otto set up a US subsidiary, *The Otto Gas Engine Works,* in Philadelphia in 1878 after exhibiting the engine at the 1876 Expo in Philadelphia. Daimler worked for Otto before setting up his own business with a subsidiary in New York City. Later Daimler made a lightweight engine for Zeppelin. (See Chap. 2.)

By the time of the Wright brothers powered flight, the gas engine and the gasoline engine had entered the '*knowledge commons* ' of workshops all over the industrial world, making it difficult to trace the influence path from Otto to the Wrights. In *Scientific American* of 1900, one can find advertisements and articles about gasoline engines from Otto, Daimler, Charter, Webster, Olin and Olds, all in the United States to mention a few. In Buffalo the E.R. Thomas Co. made engines, automobiles and motorcycles. Curtiss bought engine castings from Thomas. The development of the internal combustion engine had reached beyond the 'tipping point' or avalanche stage of technology, making it difficult to create an accurate influence network.

4.2.2 The Role of Aeronautical Science Nodes

In his seminal work *A History of Aerodynamics,* Anderson describes the lag between mathematically based fluid mechanics theory and practical advances in "heavier-than-air" machines up to the early twentieth century. Theorists such as Joukowski, Kutta and Prandtl, developed a mathematical theory of lift, for example, after dozens of machines were already in the air. The history of science however often recognizes mathematical codification of theory rather than empirical science. For example, Langley, Lilienthal, Lanchester developed theories of flight in the 1890's. Langley and Lilienthal performed extensive experiments with wind tunnel facilities and published data on lift and drag before the Wright brothers' experiments. These men were not amateurs. Langley had training at Harvard in Astronomy, Lilienthal had training at TH Berlin under Franz Reuleaux, and Lanchester studied at the Royal College of Science in London. Yet books in fluid mechanics do not often refer to the Lilienthal theory of flight or the Lanchester vortex theory of wing-lift (Anderson is the exception). The Wrights of course made extensive scientific experiments on lift, but they had the hindsight and examples of Langley and Lilienthal to follow. For recent books on the history of aeronautics and flight see Bloor (2011) and Eckert (2007).

The aviation network also makes clear the connection between aeronautical research in the nineteenth century. and the biomechanics studies of bird flight such as Marey, Pettigrew and Mouillard. Mathematical theory took a back seat in the nineteenth century vis a vis aeronautics, even though giants such as Navier, Stokes, Reynolds, and Helmholtz were mathematically codifying fluid mechanics. Helmholtz is even quoted as pronouncing that heavier-than-air flying machines were not possible.

When historians write a history focusing on one inventor, the network described can be often characterized as a 'star' network, with a central node and radiating branches to other nodes as in the Chanute sub-network in Fig. 4.9. In order to create a network such as the one in Fig. 4.2, one needs to read the histories of many inventors and connect the separate star networks with common links. Sometimes this is done in surveys of a technology as in the work of Gibbs-Smith (1954) or Hallion (2003) . However the minor nodes and their links are not always easy to uncover. The links from Ernest Archdeacon, Chairman of the Aero Club de France for example, were important in connecting Chanute, Voisin, Bleriot and many other French aviators. Alexander Bell was a link to a half dozen innovators including Glenn Curtiss.

4.3 Ballooning Evolution as a Precursor to Aviation

One of the important precursors to heavier-than-air aviation was ballooning and the associated science of meteorology. Ballooning was one of the technologies of the late eighteenth and nineteenth centuries that did not exhibit the exponential growth of the steam engine or the internal combustion engine except in its very early stage.

The key nodes in the art and science of ballooning were:

Montgolfier brothers [Joseph M. 1740–1810, and Jacque Etienne, 1745–1799] In 1782–83 they created hot air balloons, first with animals and then with human passengers. Demonstrations of early flights inspired much interest and other successful flights.
Robert brothers [Anne-Jean 1758–1820, and Nicolas-Louis 1760–1820] Early French balloonists, were the first to use hydrogen in a balloon in 1783; with Jacque Charles, they also built a dirigible shaped steerable balloon in 1784.
Jean-Pierre F. Blanchard [1753–1809] French aeronaut, flew across the English Channel with John Jefferies in1785. Blanchard was later killed in a parachute jump from a balloon.
Charles Green [1785–1870] Early British balloonist. In 1821 he used coal gas instead of hydrogen gas and made a long distance flight of 770 km. Green made over 500 flights in his career.
John Wise [1808–1879] Early American balloonist with over 400 flights. Published an influential book in 1850, *A System of Aeronautics*.
Thaddeus S.C. Lowe [1832–1913] American balloonist who created a Union Army balloon corps for reconnaissance during the Civil War 1861–1863. Contributed to the science of meteorology. Studied the book of John Wise on balloon construction.
Paul Haenlein [1835–1905] German balloonist and mechanical engineer. He built a semi-rigid dirigible balloon with a 6 hp Lenoir-type internal combustion engine, c. 1870.
Tissandier brothers [Gaston 1843 –1909 and Albert 18 -18] Gaston was a meteorologist made an ascent to 8600 meters in 1875. Later the brothers used a Siemens electric motor in 1883 to power a dirigible-type balloon.
Alberto Santos Dumont [1873–1932] Brazilian aviator, who built an internal combustion-powered dirigible in controlled flight around the Eiffel Tower in Paris in 1901.
Count Ferdinand von Zeppelin [1838–1917] Prussian military officer sent to study the US Army Balloon Corps of Thaddeus Lowe. Later he designed large steerable rigid dirigibles using Daimler internal combustion engines c. 1900.
Thomas S. Baldwin [1860–1923] American steerable dirigible designer. Used a motor built by Glenn Curtiss for a dirigible bought by the US Army Signal Corps in 1900.

Following our method in this book, we integrate the number of timeline events as a measure of the knowledge base, as shown in Fig. 4.13. This figure shows that after

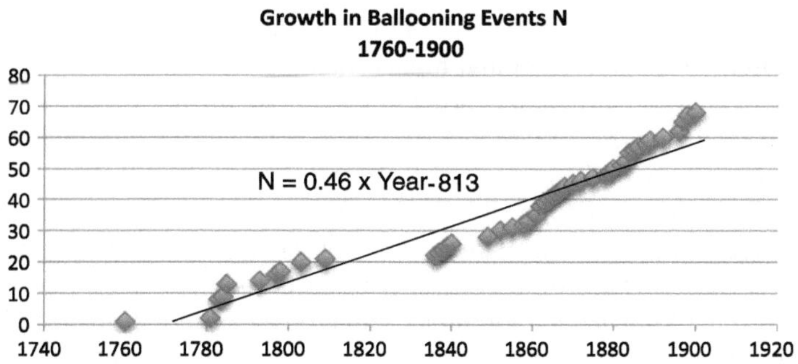

Fig. 4.13 Integration of timeline events in ballooning from Wikipedia supplemented with data from John Wise (1850) history of ballooning and Alfred Zahm 1911. Except for the first few years, this data shows a linear or multi-linear type growth. The best linear fit equation using Excel is shown in the graph

an initial jump in number of events, the growth throughout the nineteenth century, from the free motion balloons of the Montgolfier Brothers c. 1790 to the controlled flights of the dirigibles of the early Zeppelin era c. 1900, was a linear progress in integrated events or a bilinear one with a pause from1810–1840. This chart is based on a Wikipedia timeline as well as on books of the period such as John Wise (1850) and Zahn (1911). This chart may be biased toward North American ballooning but as noted in other node growth charts, one can assume that if events are missing, they are uniformly absent and their addition would not change the overall qualitative linear shape of the growth curve from 1850–1900.

By the beginning of the twentieth century, ballooning technology became important for early twentieth century. aviation. After a century of ballooning, the societies of Western Europe and North America were acclimated to the idea of human flight. Ballooning was also the inspiration of aerial navigation societies such as the Societie Aerostatique et Meteorology de France founded in 1852. These societies had thousands of members world-wide who either took an active interest in ballooning or were themselves pilots or passengers. By the time of Bleriot and the Wright Brothers, ballooning technology had created a cadre of designers and technicians in the manufacture and maintenance of ballooning aircraft. Similar to airplanes in the early twentieth century, in the nineteenth century, balloons were used for exhibitions and competitions and by the turn of the century, hundreds of thousands of people had the experience of seeing flying machines aloft. The idea of using engines, propellers, and control surfaces in balloon craft was common by 1900 as seen in the machines of Santos Dumont and Count Zeppelin. Also, instruments to measure the altitude and properties of the atmosphere were in the knowledge commons of aviation at the beginning of the Wright Brothers era.. Even parachute technology was known at this time.

Of particular importance to twentieth century aviation was the science of meteorology that was advanced with the technology of ballooning in the nineteenth century. Balloons mapped out vast regions of the atmosphere. Whereas heavier-than-air machines circa 1910 rarely ventured higher than 10,000 ft or 3300 m, balloons had ascended to over 7000 m a century earlier in 1803. Also the fuel capacity of airplanes limited their range to 100–200 km in 1910, while balloons and dirigibles had logged over 1000 km in early flights.

Thus while balloons and dirigibles of the nineteenth century did not exhibit the exponential growth of the twentieth century airplane, they laid the foundation for the explosive development of aircraft in the first decade of the new century. They contributed to the knowledge 'commons' in aeronautics in the nineteenth century. Ballooning can be thought of as a necessary part of the evolution of human flight, providing key scientific and technical knowledge that eventually fed the self-sustaining growth of early aviation. Using the avalanche or sand pile analogy, the linear growth of knowledge in ballooning is like the build-up of a sand pile and the exponential development of aviation at the end of this period is like the critical state of the sand pile in which small events, such as the lightweight internal combustion engine, trigger the avalanche of innovation that resulted in human flight.

4.4 Summary

In his seminal history of the steam engine, Robert H. Thurston, (1878) the first President of the American Society of Mechanical Engineers, wrote; "*Great inventions are never, and great discoveries are seldom, the work of any one mind.* Every great invention is really the aggregation of minor inventions or the final step of a progression." Aviation is certainly a field that saw a large number of steps on the way to human flight from Langley's powered model aircraft, not unlike the drones of modern warfare. Lilienthal gave his life to conduct experiments with human glider flight. Octave Chanute finished Lilienthal's work by the development of the Pratt truss biplane topology that Orville and Wilbur Wright used. And of course powered flight would not have been powered without the advances of Lenoir, Otto and Daimler.

In this chapter we have presented evidence for a social network theory of invention in early aviation primarily based on historical books and documents from the period 1890–1910. This model makes explicit what earlier historians have posited on the creation of heavier than air machines; namely that the seeds for human flight were laid in the mid to late nineteenth century and grew out of the contributions of many artisans, engineers, fluid scientists, industrialists as well as educational and professional institutions. The network model makes explicit the contributions of earlier studies of bird flight as well as the critical development of the internal combustion engines. This model also highlights the importance of communication and transportation technologies in the globalization of the effort to create heavier than air flying machines through publications and international meetings. The statistics of the node-links in the early aviation network appear to have anticipated those of modern twenty-first century networks.

In this study there are other minor and major nodes and links that are not included here but are listed in Chanute's book. However, further study will only add more nodes and links to the already complex patterns of Figs. 4.2 and 4.9. It can only serve to confirm the truth that human innovation in aviation was the result of collective action and communication, rather than the product of lone 'geniuses' working in isolation.

When one reads a biography of the Wright brothers, there is often a sense of their working in secrecy, the desolate dunes of the Kitty Hawk camp in the summers of 1902, 1903. Afterwards we read of the attempts to keep the details of the wing warping under wraps and later the use of patent law to prevent others from creating similar flying machines. But when one looks at the milieu of aeronautics in the two decades of the 1890's and 1900's there is a contrasting sense of openness, of humans desperately seeking the new, the adventurous, some for the satisfaction of knowledge like Langley and others like Santos Dumont seeking the thrill of flying. And in the midst of all these personalities is the growth of an aviation *'knowledge commons'* that is seeping into Western culture through its magazines, newspapers and books, telling the world about Lilienthal, Maxim, Wenham and others, exaggerating their feats, questioning their accomplishments and stirring the imagination of young boys in Dayton, Ohio.

Acknowledgments

Appendix I

Human Nodes 1800–1910 in Aviation Network

Clemant Ader [1841–1925]; Alessandro Anzini [1877–1956]; Ernest Archdeacon [1863–1957]; William Avery; Thomas Baldwin [1860–1923]; Bedell; Alexander Bell [1847–1922]; Karl Benz [1844–1929]; Louis Bleriot [1872–1936]; Louis Breguet [1880–1955]; George Bryan [1864–1928]; George Cayley [1773–1857]; Octave Chanute [1832–1910]; Glenn Curtiss [1878–1930]; Gottlieb Daimler; Geoffrey deHavilland [1882–1965]; Leon Delagrange [1873–1910]; William Durand; Jacob Ellehammer [1871–1946]; Henry Farman [1874–1958]; Ferdinand Ferber [1862–1909]; Hans Grade [1879–1946]; Claude Grahame-White [1879–1959]; Lawrence Hargrave [1850–1915]; Helmholtz; William Henson [1812–1888]; Edward Huffaker [1856–1936]; Augustus Herring [1867–1927];J. Frederick Jane [1865–1916]; Nikolai Joukowski [1847–1921]; Hugo Junkers [18559–1935]; Martin Kutta [1867–1944]; Frank Lahm [1877–1963]; Frederick Lanchester [1868–1946]; Eugen Langen [1833–1895]; Samuel Langley [1834–1906]; Lenoir; Leonardo da Vinci [1452–1519]; Leon Levavasseur [1863–1922]; Otto Lilienthal [1848–1896]; Thaddeus Lowe [1832–1913]; Charles Manly [1876–1927]; Etienne Marey [1830–1904]; Marks; Marriott; Martin; Hiram Maxim [1840–1916]; Wilhelm Maybach [1846–1929]; James Means; Joseph and Jacques de Montgolfier [1740–1810, 1745–

1799]; John Montgomery [1858–1911]; Louis P. Mouillard [1834–1897]; Alexander Mozhaiski [1825–1890]; Nikolas Otto [1832–1891]; Alphonse Penaud [1850–1880]; James Pettigrew [1832–1908]; Horatio Phillips [1845–1926]; Percy Pilcher [1867–1899]; Ludwig Prandtl [1875–1953]; Franz Reuleaux [1829–1905]; Ferdinand Redtenbacher; Charles Rolls [1877–1910]; Alberto Santos-Dumont [1873–1932]; Thomas Selfridge [1882–1908]; Elmer Sperry [1860–1930]; Spratt; John Stringfellow [1799–1883]; Victor Tatin [1843–1913]; Edwin Thomas; Robert Thurston [1839–1903]; Gabriel Voisin [1880–1973]; Theodore von Karman [1881–1963]; Orville and Wilbur Wright [1871–1948, 1867–1912]; Francis Wenham [1824–1908]; Gustav Whitehead [1874–1927]; Albert Zahm [1862–1954]; Ferdinand von Zeppelin [1838–1917];

Source Nodes versus Unidirectional Links

Wright Bros.	k = 24
Octave Chanute	k = 23
Samuel Langley	k = 15
Otto Lilienthal	k = 16
Glenn Curtiss	k = 11
George Cayley	k = 10
Franz Reuleaux	k = 10
Robert Thurston	k = 10
Alexander Bell	k = 9
Lawrence Hargrave	k = 9
John Montgomery	k = 9
Percy Pilcher	k = 9
John Stringfellow	k = 8

Appendix II

Aeronautical Societies Nodes

1866 The Aeronautical Society of Great Britain [120 members in 1906]
1872 Societe Francais de Navigation Aerienne [103 members in 1906]
1880 Russian Aeronautical Society
1881 Berliner Verein Fur Luftschiffahrt [807 members in 1906]
1887 Flugtechnischer Verein in Wein [90 members in 1906]
1889 Munchener Verein fur Luftschiffahrt [383 members in 1906]
1886 The International Commission for Scientific Aeronautics [Paris]
1897 Aeronautical Club de France [500 members in 1906]
1898 Aero-Club, Societe d'encouragement a la locomotion aerienne
1900 The Permanent International Aeronautical Committee [33 elected members; included O. Chanute]
1900 Svenska Aeronautiska Sallskapet [80 members in 1906]
1901 Aero-Club de Belgique [300 members in 1906]
1901 Schweizerischer Aero-Club [140 members in 1906]
1902 Deutscher Luftschiffahrt-Verband

1902 The Aero-Club of the United Kingdom- A society for the encouragement of aerial locomotion [700 members in 1906]
1904 Societa Aeronautica Italiana [182 members in 1906]
1905 Federation Aeronautique Internationale [Paris]
1905 The Aero-Club of America [275 members in 1906]
1905 El Real Aero-Club de Espana [105 members in 1906]
1910 Intercollegiate Soaring Clubs [USA]

Source Moedebeck (1906) *The Pocket Book of Aeronautics* pp 436-444.

Appendix III

University Nodes in the Early Aviation Network

Technische Hochschule Berlin [Lilienthal, Junkers, Spade]
Cornell University [Chanute Lectures, Zahm, Manly, Sperry, Grumman]
Santa Clara College CA [Montgomery]
Royal College of Science (Now Imperial College) [Lanchester]
Notre Dame University [Zahm]
Catholic University of America [Zahm wind tunnel]
Gottingen University [Prandtl wind tunnel]
Allegheny College [Langley]
MIT [Lionel Mark, Aircraft Engines]
Stanford University [William Durand, propeller lab]
Columbia University [first aeronautical engineering degree]
US Army Schools of Military Aeronautics [Cornell, Princeton, Ohio State, Texas]
Intercollegiate Soaring Association [Cornell, Yale, Princeton, Columbia, c. 1910]
US Naval Academy[Langley, Thurston]
Moscow University [Joukowski]
Munich University [Kutta]

Appendix IV

Sibling Nodes

Breguet , Louis and Jacques
Farman, Marurice A. and Henry
Lanchester, Frederick W. and George
Lilienthal, Otto and Gustavus
Lougheed, Alan and Malcolm
Moisant, John, Alfred, and Matilde
De Montgolfier, Joseph-Michel and Jacques-Etienne

Percy and Ella Pilcher
Stinson, Majorie, Katherine and Eddie
Voisin, Gabriel and Charles
Wright, Orville and Wilbur

Source Longyard, W.H. (1994) Who's Who in Aviation History: 500 Biographies.

References

Abbot, C.G. (1932) Great inventions. Smithsonian Scientific Series, vol 12, The airplane, pp 225–239

Anderson JD Jr (2004) Inventing flight: the Wright brothers and their predecessors. John Hopkins Press, Baltimore

Anderson JD Jr (1997) A history of aeronautics and its impact on flying machines. Cambridge University Press, Cambridge

Barabasi A-L (2003) Linked. Plume-Penguin Publishers, Cambridge

Bloor D (2011) The enigma of the aerofoil: rival theories in aerodynamics 1909–1930. University of Chicago Press, Chicago

Cayley G (1809–1810) On aerial navigation. Nicholson's J (November 1809, February 1810, March 1810)

Chanute O (1896) Progress in flying machines. Reprinted by Dover Publications in 1997

Crouch T (1981) A dream of wings. Norton, New York

Crouch T (1989) The bishop's boys: a life of wilbur and orville wright. W W Norton & Co., New York

Curtiss GH, Post A (1912) The curtiss aviation book. Frederick A. Stokes Co, New York

Eckert M (2007) The dawn of fluid mechanics: a discipline between science and technology. Wiley, New York

Gibbs-Smith CH (1954) A history of flying. Frederick A Praeger, New York

Graham-White C, Harper H (1911) The aeroplane, past, present and future. J.B. Lippincott Co., Philadelphia

Hallion RP (2003) Taking flight. Oxford Press, New York

Hatch A (1942) Glenn Curtiss: pioneer of aviation. The Lyon's Press, Guilford

Herring-Curtiss archive, Cornell University

Jane, F. (1917) Jane's historical aircraft, Facsimile of 1917 printing

Lanchester FW (1907) Aerodynamics. A. Constable & Co., London

Langley S (1911) Langley memoir on mechanical flight. In: Manly C (eds), Smithsonian institution, Washington

Longyard WH (1994) Who's who in aviation history: 500 biographies. Presidio Press, Novato

Lougheed V (1909) Vehicles of the air. The Reilly and Britton Co., Chicago

Lilienthal O (1889) Bird flight as the basis of aviation: a contribution towards a system of aviation, Reprinted 2001, in translation from the German, by American Aeronautical Archives

MacFarland MW (1953) The papers of Wilbur and Orville Wright: including the chanute—Wright letters and other papers of octave chanute, vol 1, 2. McGraw-Hill Book Co, New York

Mackersey I (2003) The Wright brothers. Oxford Press, New York

Magoun FA, Hodgins E (1931) A history of aircraft, Whittlesey house. McGraw-Hill Book Co, New York

Marey EJ (1874) Animal mechanism. D. Appleton & Co, New York

Maxim H (1908) Artificial and natural flight. Whittaker & Co, London

Means J (1896) The aeronautical annual. W. B. Clarke Publisher, Boston

Means JH (1964) James means and the problem of manflight. Smithsonian Institution, Washington

Milbank J Jr (1943) The first century of flight in America. Princeton University Press, Princeton
Millbrooke A (2006) Aviation history. Jeppesen Press, Englewood
Moebedeck (1906) Pocket book of aeronautics
Moon FC (2012) A social network model for innovation in early aviation history. In: Koetsier T, Ceccarelli M (eds) Explorations in the history of machines and mechanisms, pp 3–19. Springer, Dordrecht
Mouillard L (1881) Empire de l'Air, [The Empire of the Air]
Newman MEJ (2003) The structure and function of complex networks. SIAM review
Short S (2011) Locomotive to aeromotive: Octave Chanute and the transportation revolution. University of Illinois Press, Illinois
Shulman S (2002) Unlocking the sky: Glenn Curtiss and the race to invent the airplane. Harper-Collins Publishers, New York
Sibley J Engineering (1993–1903), The Crank (1886–1892), Sibley College, Cornell University
Thurston RH (1878) A history of the growth of the steam engine. D. Appleton and Co., New York
Von Karman T (1954) Aerodynamics: selected topics in the light of their historical development. Cornell University Press, Ithaca
Watts DJ, Strogatz SH (1998) Collective dynamics of small world networks. Nature 393:440–442
Watts DJ (2003) Six degrees. Norton, New York
Wise J (1850) A system of aeronautics. Joseph A Speel, Philadelphia
Zahm AF (1911) Aerial navigation. D. Appleton and Co, New York

Chapter 5
Wireless and Radio Electronics Social Networks

Abstract The seeds of wireless telegraphy grew out of the network surrounding the magnetic-mechanical telegraph. In addition to Marconi, Deforest and Armstrong, there was a complex set of up to fifty players spanning the work of Maxwell (1865) to the early days of radio and RCA's David Sarnoff in the 1920s. This technology network has links to modern Chaos Theory in the late twentieth century [Chap. 6]

The evolution of radio communication grew seamlessly from mechanical telegraphy in the late eighteenth and early nineteenth centuries. The original mechanical-optical telegraph systems of the late eighteenth century by the Chappe brothers were like mechanical semaphore stations installed atop of churches and towers in France and Germany. (Standage 1998) When Samuel Morse [1838] and William Cooke and Charles Wheatstone [1839] invented a system based on electromagnetism around 1840, it was called the 'magnetic' telegraph to distinguish it from the mechanical signal systems used throughout Western Europe. (Silverman 2003; Standage 1998) The transatlantic telegraph between Europe and North America began operation in 1866.

With the discovery of electromagnetic waves by James C. Maxwell and Heinrich Hertz, inventors such as Guglielmo Marconi around 1900 tried to replace the magnetic telegraph, which used copper wires, with a telegraph system without wires. Hence the term 'wireless' telegraph. The wireless telegraph, like its predecessor the magnetic telegraph, was a binary digital communication system; the so-called dots and dashes or long and short pulse system. The sequence of dots and dashes, such as the Morse Code, similar to our zeroes and ones today in digital electronics, was coded to map the sequence into human language.

The basic components of the wireless telegraph system were an electromagnetic wave generator, a transmitting antenna, a receiving antenna and receiving electrical circuit to record the short and long pulses. Originally the transmitter was based on an electrical spark between two electrodes. Very quickly inventors discovered that the system could be improved by placing the electrodes in a vacuum tube. Circuits for controlling the wavelength of the electromagnetic waves were then invented

F. C. Moon, *Social Networks in the History of Innovation and Invention*,
History of Mechanism and Machine Science 22, DOI: 10.1007/978-94-007-7528-2_5,

© Springer Science+Business Media Dordrecht 2014

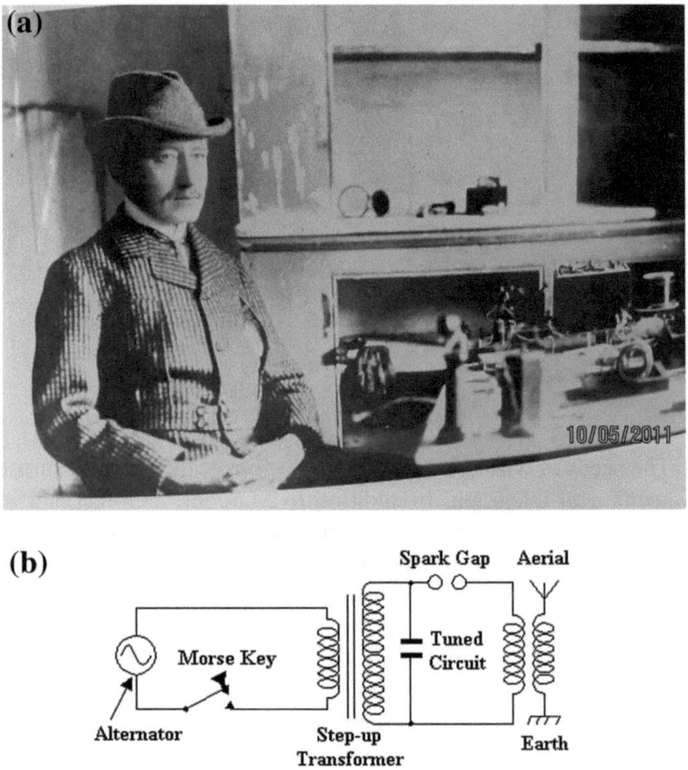

Fig. 5.1 **a** Guglielmo Marconi [1901] with electronics equipment for the first trans-Atlantic transmission. [Courtesy Columbia University Archives.] **b** Marconi type transmitter circuit. (From History of Radio: Bev Parker, http://www.localhistory.scit.wlv.ac.uk/Museum/Engineering/Electronics/history/earlytxrx.htm) See also Morgan (1913)

and resonant circuits for receiving were discovered. Lee de Forest added another electrode to the vacuum tube that allowed it to detect several frequencies at once, as are present in music and voice, and the radio was born. Edwin Armstrong invented highly sensitive circuits so that the radios could fit into the average American home. The final step in the system was to create an audience of listeners and to entice them to purchase radio receivers to listen to the voice and music broadcast over the electromagnetic spectrum. David Sarnoff who originally worked for Marconi's wireless telegraph was a key player in the formation of the Radio Corporation of America or RCA in the 1920s. (Lewis 1991) For a popular history of Marconi and wireless technology see the recent book of Erik Larson (2006).

The evolution of wireless telegraphy had its roots in the networks of the magnetic telegraph and electrical technology of the nineteenth century, including the electric light bulb. However we shall focus our attention on the period, 1880–1920 and nodes associated with electromagnetic wave communication. All the nodes in the wireless/radio technology can be connected with a set of links in a historical social network.

The principal nodes in the science and technology of wireless and radio were:

James Clerk Maxwell [1831–1879] Organized the equations of electromagnetics and discovered the wave nature of electricity.
Heinrich Hertz [1857–1894] Experimentally demonstrated the existence of radio waves predicted by Maxwell.
Oliver Lodge [1851–1940] British; early research on wireless principles.
Guglielmo Marconi [1874–1937] Brought together many elements to create a working wireless system.
Edouard Branly [1844–1940] Invented a detector of radio waves, called a coherer.
Karl F. Braun [1850–1918] Invented crystal detector of radio waves, invented tuned circuit for transmission of radio waves.
Ian Fleming [1849–1945] Physicist consultant to the Marconi Company, invented a vacuum tube.
Reginald A. Fessenden [1866–1932] Canadian engineer who broadcast the first words using AM modulation.
Lee de Forest [1873–1961] Created the triode vacuum tube to amplify radio waves.
Edwin H Armstrong [1890–1954] Invented regenerative electronic circuits with extreme sensitivity.
David Sarnoff [1891–1971] A Marconi wireless operator who later helped create and run the RCA radio network.

5.1 Timeline Avalanche

A partial timeline of events in the evolution of wireless and radio is shown in Fig. 5.2, beginning with early discoveries in electric and magnetic physics by Oersted and Faraday, followed by the development of the magnetic telegraph by Morse and others. The discovery of the wave nature of electromagnetic fields by Maxwell and Hertz are key enabling events in this timeline. Work by the British Physicist Oliver Lodge, the German Karl Braun and Italian-Irish Marconi followed near the beginning of the twenty-first century. Lee de Forest's audion vacuum tube and Armstrong's regenerative circuits followed in the next two decades. It is easy to assemble more than sixty events from 1800–1920 (Lewis 1991).

If we graph the sum of the number of major events, we obtain an exponentially increasing curve shown in Fig. 5.3, similar to the event timeline curve of early aviation in Chap. 4 (Fig. 4a). The exponential growth suggests that as new discoveries are made, more players are inspired to contribute to the science and technology. The growth is not unlike the spread of a disease, except it is the spread of ideas that are converted into artifacts we call inventions.

The graph in Fig. 5.3 appears to have an inflection point near 1910 and might be best fit with an 'S' curve. However we would have to include events after 1920 before we could say that the growth was slowing down. This presentation of historical data

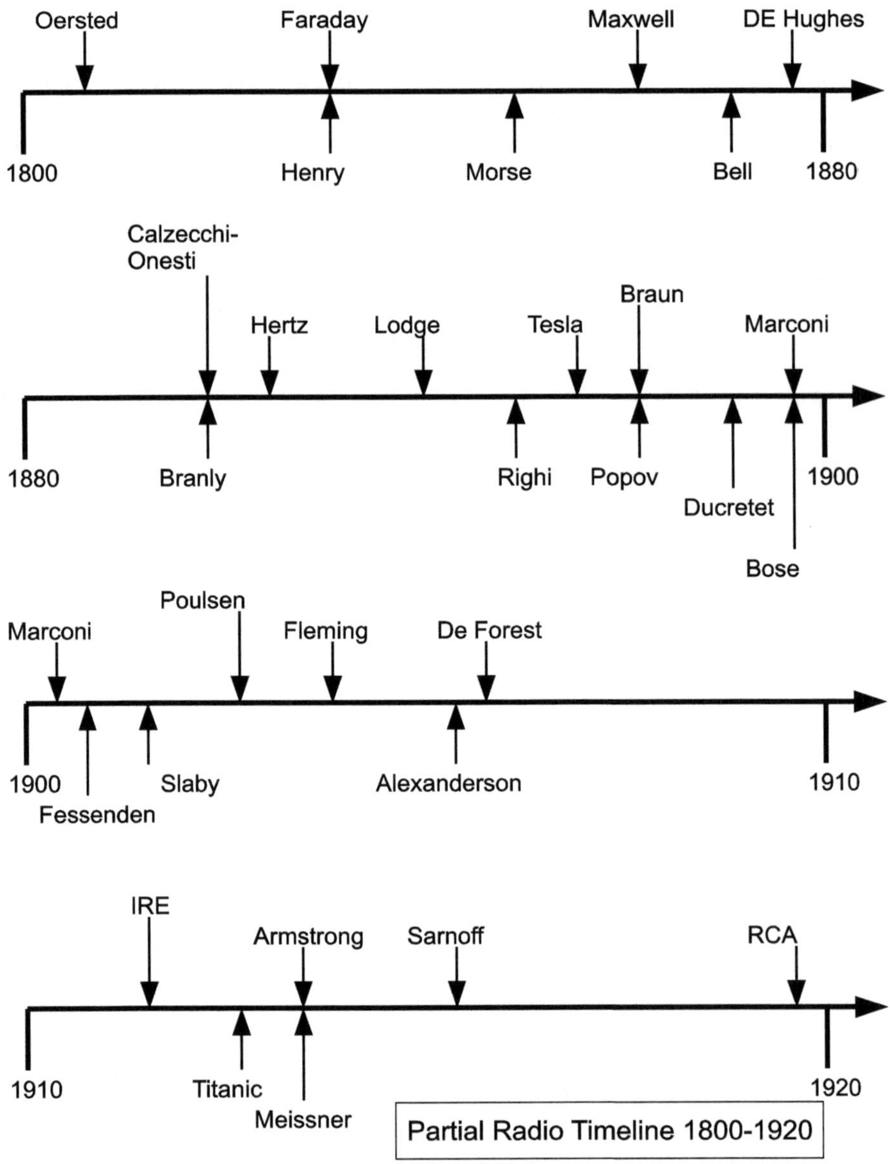

Fig. 5.2 Partial wireless and radio electronics timeline 1800–1920 showing approximately half of the significant events and nodes captured in the integration of events in Fig. 5b. For sources see References in this Chapter

supports two theses of this book, namely the evolutionary nature of innovation and the creation of social and institutional networks that produce this evolution.

The historical data upon which this curve is based has been gathered from several sources including books, archival material and the web. Different sources have their

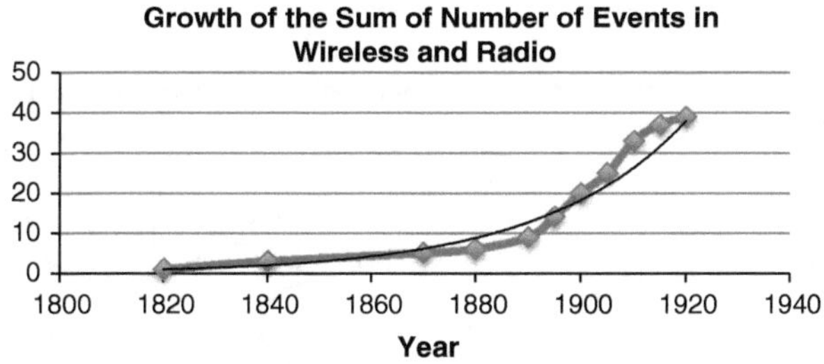

Fig. 5.3 Growth of the number of events in the history wireless-radio electronics in Fig. 5.2. The *solid line* is a best-fit exponential curve

own view of history. For example WikiPedia has several websites for 'Wireless and Radio History' There is overlap of course but different authors include or exclude many minor events and nodes such as Alexander Popov of Russia. In most cases the key engineers, technicians and artisans are almost always forgotten even though artisan skill can play an essential role in a new technology. For example, in the case of the de Forest vacuum tube, the glass artifacts were made by a well-known glassmaker Henry W. McCandless of New York who also made headlamps for automobiles (Lewis 1991).

In some examples of ' radio timelines' the history of 1800–1950 is almost non-existent such as in the "Wireless History Foundation" This organization only documents the modern wireless age of cell phones, beginning around 1983. Their timeline for pre-1900 only mentions Marconi and there is no mention of anything historic from 1900–1927. The IEEE [Institute of Electrical and Electronic Engineers] Timeline on their website is only a little better.

Thus the integrated innovation events in Fig. 5.3, must be considered a lower bound on the explosion of activity in wireless and radio electronics beginning in the period around 1880–1885. The period 1800–1880 saw a steady but linear accumulation of technical and scientific knowledge. It is the thesis of this book that the exponential rise in the accumulated knowledge related to radio was preceded by a linear accumulation of knowledge and experience and was the result of a historical social network of individuals and institutions as shown in Fig. 5.4. We don't know the critical size of this innovation network for which exponential growth occurred. But clearly the evidence is there to show it took place. We don't know if this sudden growth in knowledge had the characteristics similar to sandpile or snow avalanches as described in the so-called "Self-organized Criticality" theory of Bak, but the many examples in this book suggest that scientific and technical innovation might be candidates for SOC modeling.

The biographies of each of the principal players provide links to other players. In the case of Edwin Armstrong , we used archival records at Columbia University to construct a "star network" below. When these so-called star networks are put together, they form a network of around 40 nodes shown in Fig. 5.4.

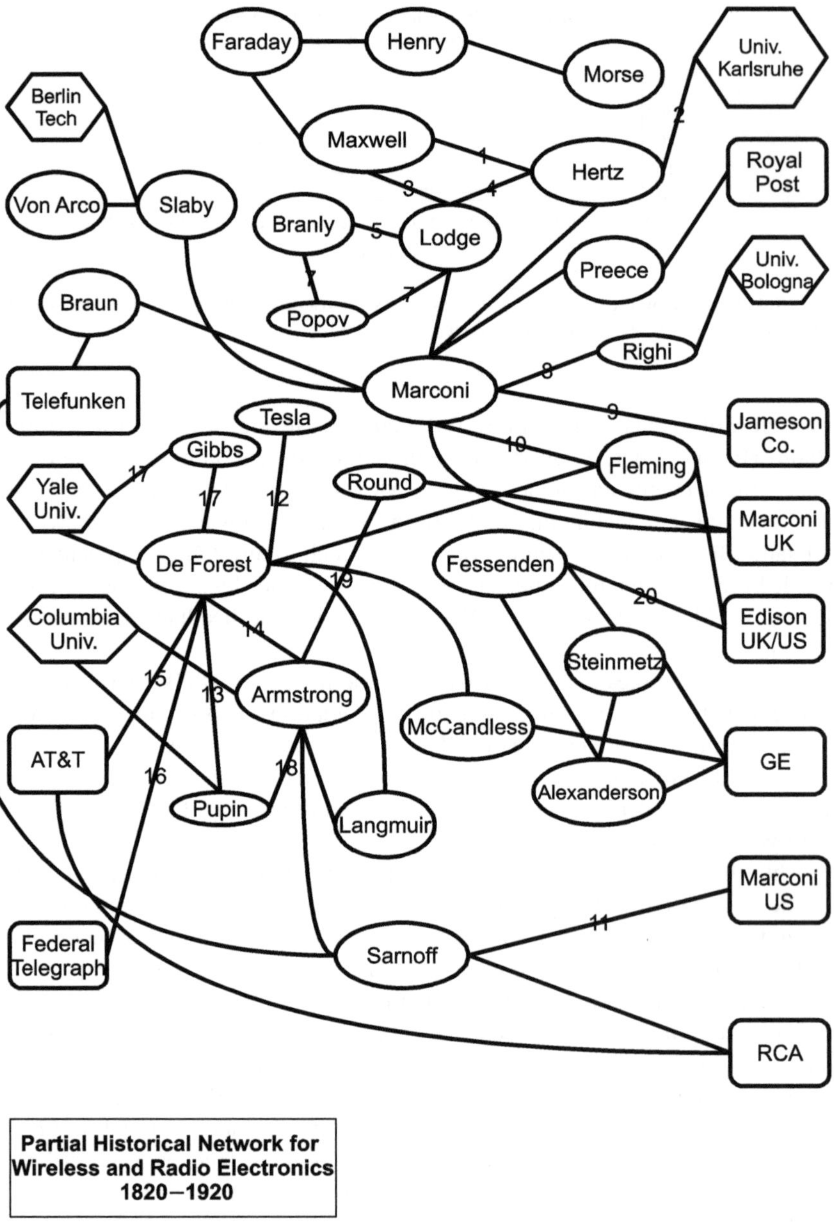

Fig. 5.4 Partial historical innovation network for wireless and early radio technology 1820–1920. Numbers on links refer to table of references below

5.2 Selected Links in the Radio Electronics Historical Innovation Network [See Fig. 5-4]

1. Heinrich Hertz [1857–1894] worked on a proof of Maxwell's theory of electromagnetic waves while working at the University of Berlin with Professor Harmann von Helmholtz [1880]. [Ref. 1, see end of list]
2. Hertz while at the University of Karlruhe used a 'Leyden jar' to create sparks and oscillating currents in spiral coils. [1886] [Ref. 1]
3. Sir Oliver Lodge [1851–1940] heard Maxwell lecture at University College London, [1873]. [Ref. 1]
4. Lodge and Hertz corresponded and exchanged scientific papers. [Ref. 2]
5. Lodge worked and improved the Branly [1846–1940] coherer in 1840. [Ref.2]
6. Dr Alexander Muirhead (Telegraph engineer) heard a lecture by Oliver Lodge at the Royal Institution. [1894] [Ref.1]
7. Alexandr Stepanovich Popov [1859–1906] repeated the experiments of Hertz and Lodge using a Branly-Lodge coherer. [1895] [Ref.1]
8. Guglielmo Marconi [1874–1937], was an informal student at the University of Bologna and heard physics lectures about the work of Faraday, Maxwell and Hertz. [Ref.3]
9. Marconi's Anglo-Irish mother Anne Jamieson was linked to the Jamieson Company that produced alcohol products in the UK and helped fund Marconi's UK wireless telegraph company. [Ref.3]
10. Marconi hired John A Fleming from the Edison Company in London. Fleming had studied under Maxwell. [Ref.3]
11. The Marconi Company in New York hired David Sarnoff as a telegraph operator. Sarnoff met Marconi during a trip to the US. [c. 1906] [Ref.3]
12. Lee de Forest [1873–1961] attended a lecture by Nicoli Tesla while at Yale U. that inspired his study of electricity.[Ref. 3]
13. Lee de Forest while at Yale knew of the work of Columbia University Professor Pupin who later mentored Armstrong. [Ref.3]
14. Lee de Forest met Armstrong during a lecture that the latter gave at Columbia University. [1913] [Ref.3]
15. De Forest sold his 'audion' rights to AT&T for $50,000 [1913] [Ref.3]
16. De Forest worked for the Federal Telegraph Co. a competitor of the Marconi Co. [1910], [Ref.3]
17. De Forest studied mathematics and electromagnetic theory with J Willard Gibbs of Yale University. [Ref.3]
18. Armstrong attended the Columbia University School of Mines, Engineering and Chemistry and entered the Electrical Department under Prof Michael Pupin. [Ref.3]
19. Armstrong met Henry J. Round during WWI; the latter had worked for the London Marconi Company. [1917]. [Ref.3]
20. Fessenden was a chemist for Thomas Edison. [Ref.3]

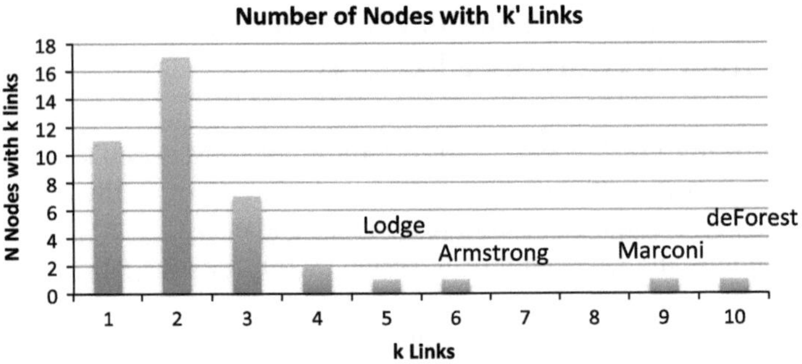

Fig. 5.5 Distribution of number of nodes with 'k' links for the network of Fig. 5.4. The high-link nodes from the right to left correspond to de Forest [10 links], Marconi [9], Armstrong [6] and Lodge [5]

{Reference 1: "Six Great Pioneers of Wireless" EBU Technical Review, Spring 1995.

Reference 2: Web on-line biography

Reference 3: *Empire of the Air,* T. Lewis (1991).}

As in the aviation network in Chap. 4, one can plot the distribution of nodes with number of links as shown in Fig. 5.5.

Thus the principal innovators have the highest number of links i.e., De Forest [10 links] and Marconi [9 links]. Armstrong has six links and Lodge five links. Sarnoff and Hertz have four links, while Alexanderson, Fessenden, Fleming, Maxwell, Pupin , Slaby and Steinmetz have three links each. Most of the other nodes in the network have only one or two links. It is likely that further historical research will reveal more complexity in the node-link pattern. This distribution is similar to that of the steam engine and aviation networks that have characteristics of power law functions found in modern social networks such as the World-Wide Web.

5.3 The Edwin Armstrong Sub-Network

It is interesting to focus on the links surrounding one or two major nodes. In the case of Edwin Armstrong, his personal and research papers are available to study at Columbia University. Also Tom Lewis's *Empire of the Air, (1991)*, devotes almost a chapter to Armstrong's career and life. The Armstrong sub-network is shown in Fig. 5.6. Since we are mainly focusing on one central node, the network has the properties of a 'star' network. The innovation network of Fig. 5.4 is the integration of many star networks.

Edwin Armstrong [1890–1954] was born after the major scientific discoveries of wireless telegraphy were in place. As a boy growing up in the suburbs of New York City, he read with fascination stories about great inventors, especially Marconi. He had a mentor near his home, Charles Underhill, who encouraged him in his

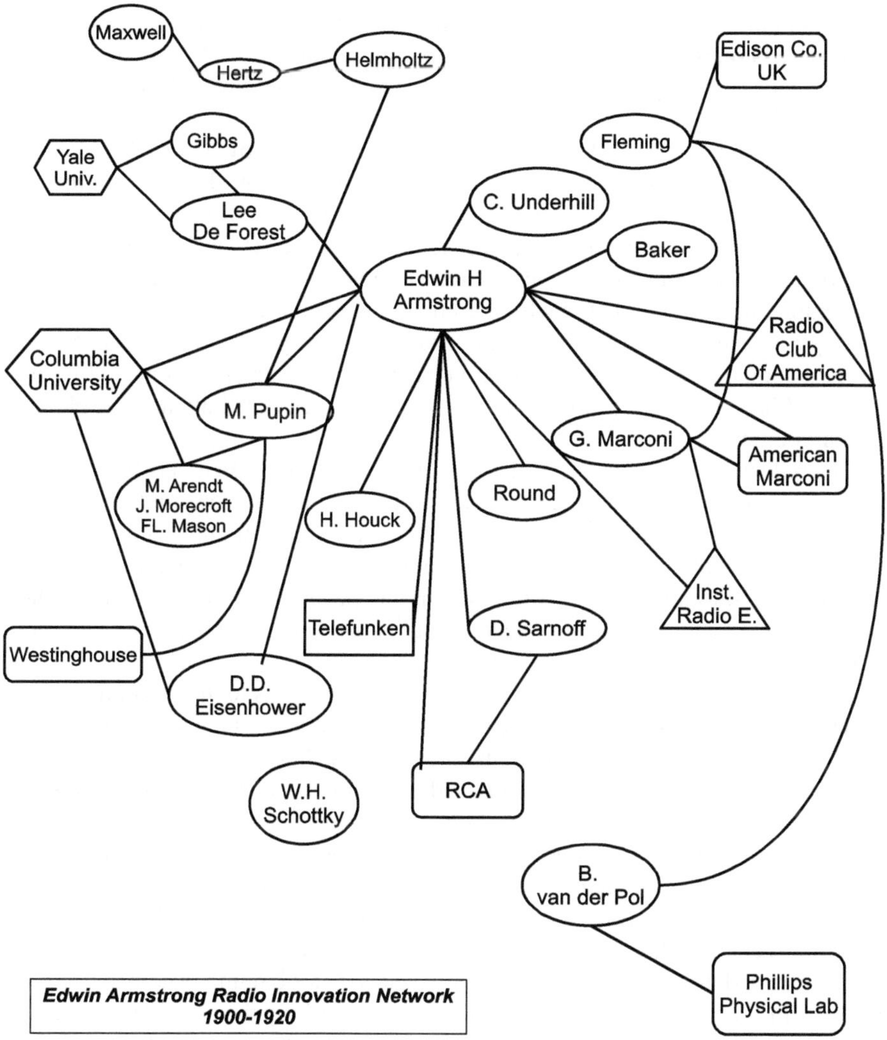

Fig. 5.6 Edwin H Armstrong radio innovation network, as a sub network of Fig. 5.4

amateur wireless activities and even gave him a de Forest audion triode to use in his circuits. During this period of 1905–1915 there were hundreds and eventually thousands of amateurs 'playing around' with wireless technology putting up antennae and communicating with each other in Morse Code. One of the popular clubs was the *Amateur Radio Clubs* of which Armstrong joined later after his college years.

Armstrong's three major contributions to radio technology were (i) the invention of a sensitive detector of electromagnetic waves with his regenerative circuit coupled to a de Forest triode tube [1913–1918], (ii) the discovery that this circuit could also become unstable and generate high frequency oscillations suitable for radio transmission, [1914] and (iii) frequency-modulated radio or FM in the 1930s. Our focus here is on the early years of radio up until about 1920. The use of the triode vacuum tube was very important because it moved wireless away from the coherer and crystal detection.

Although the subject of electromagnetics formulated by James C. Maxwell, describes electric and magnetic fields in 3D space, in practice, electrical engineering was moving to a so-called circuit-based design process in which magnetic fields were confined to inductive devices, electric fields confined to capacitive devices and electric currents confined to copper wires and resistive devices. The electron tube of Fleming, de Forest and others focused the 3D aspect of electromagnetics inside a small glass tube with a partial vacuum. Thus the major physics was localized. This discretization of electrical elements into resistors, capacitors, inductors and vacuum tubes had been under way since the days of Edison's electric bulb and the electrification of cities.

The reduction of design elements into discrete electronic elements had a parallel story in the kinematics of machines, described in Chap. 2. Here motion-changing systems were categorized in standard machine and kinematic mechanisms such as gears, pistons, linkages etc that led to efficiencies in design and manufacturing and resulted in thousands of new combinations of mechanisms and machines whose priority in the patent system produced court battles similar to those in the wireless-radio era of de Forest and Armstrong.

Armstrong's major contribution to radio electronics was to discover an optimum arrangement of these discrete electrical elements to produce a super-sensitive detector of electromagnetic signals. Hundreds and perhaps thousands of other engineers, physicists and amateurs were also engaged in a similar search. Although Armstrong was clever enough to discover the right combination of elements, if he had pursued other interests in 1910, someone in North America or Europe would have discovered regenerative circuits by the end of the First World War. De Forest himself claimed he had discovered the properties of regenerative circuits with a triode a year earlier. Some have claimed that the German physicist W.H. Schottky, of whom an electronic diode is named, had discovered regenerative electronics circuits. Another claim was made by the American rocket scientist Robert Goddard. By the year 1917, there were more than 6000 members of amateur radio clubs in North America all searching for the optimum detector circuits.

These observations do not take anything away from Edwin Armstrong or the fact that many, including his professors at Columbia University, thought him a genius. But in the technology innovation avalanche, captured in Fig. 5.3, for the wireless and radio age, many participants had access to the latest scientific and technical data and were communicating with each other sometimes daily and hourly through the very medium of wireless. (Not unlike the 'texting' today.) It was highly probable that someone would have come up with the regenerative circuit concept either by deduction or by the accident of experimentation.

At the bottom right of the Armstrong network in Fig. 5.6 is the name of the Dutch engineer-mathematician Balthazar van der Pol who worked for the Dutch radio electronics company Phillips. He had worked with Fleming in the UK before coming to Phillips. Van der Pol is included because he was one of the first to understand the mathematical dynamics of the nonlinear electronics in Armstrong's circuits with de Forest's triode tube. Van der Pol is an important node in the Nonlinear Dynamics and Chaos innovation network. We discuss his innovation network in greater detail in the next chapter.

5.3.1 Observations About the Radio Innovation Network

On the one hand the mathematical theory of electromagnetic (EM) waves developed by Maxwell and others in the early nineteenth century played a crucial role in the development of radio technology. Not only did this theory challenge Hertz to try and observe these waves, but Marconi's informal mentor, Professor Auguste Righi in Bologna, was motivated to study the physics of EM waves. On the other hand the physics of the spark or arc generator was not understood, nor were the vacuum tube devices such as the coherer and audion understood as electronic elements.

The electronic circuits used in early wireless and radio were highly nonlinear systems and the complete dynamical behavior would not be understood for a half century. (See e.g. Fig. 5.7) In the early decades a Dutch engineer for Phillips named Balthazar van der Pol began to study the dynamics of nonlinear electrical circuits and observed unexplainable erratic signals, what today we would call chaotic signals. Thus experimentalists such as Marconi, de Forest and Armstrong played roles in the invention evolution similar to that of artisans in the Renaissance machine age. For example de Forest relied on the McCandless company to produce custom made vacuum devices. One source credits a colleague of McCandless, John Grogan, with shaping the third electrode [grid] of de Forest's triode that made it into a functioning electronic component. Radio technology did not derive directly from Maxwell's mathematical equations as some physics textbooks would have one believe. There was a lot of trial and error, with many false starts and dead ends. The major players used not only ideas from physics, but also intuition, perseverance, hard work, drive, vision and entrepreneurship to produce successful products.

Part of the history of wireless and radio stems from the development of the vacuum tube, that is tied to the invention of the incandescent light bulb. There were over 20 other inventors of the light bulb from 1800 to 1879, the date of Edison's patent filing. An earlier patent by Joseph Swan in the UK was filed in 1878. In seeking to improve the light bulb, Edison had discovered thermionic emission from a heated element known as the Edison effect. John Fleming took this idea and invented an electric vacuum tube diode that could convert alternating current into direct current. This proved useful in early wireless systems. Fleming had worked for the Edison-Swan Electric Light Co in London and was later a consultant for Marconi. Thomas Edison had direct links to the magnetic land telegraph network as his first job was as a telegraph operator.

The other observation about early wireless technology is that it depended on an existing mature telegraph system that had already developed a market for communi-

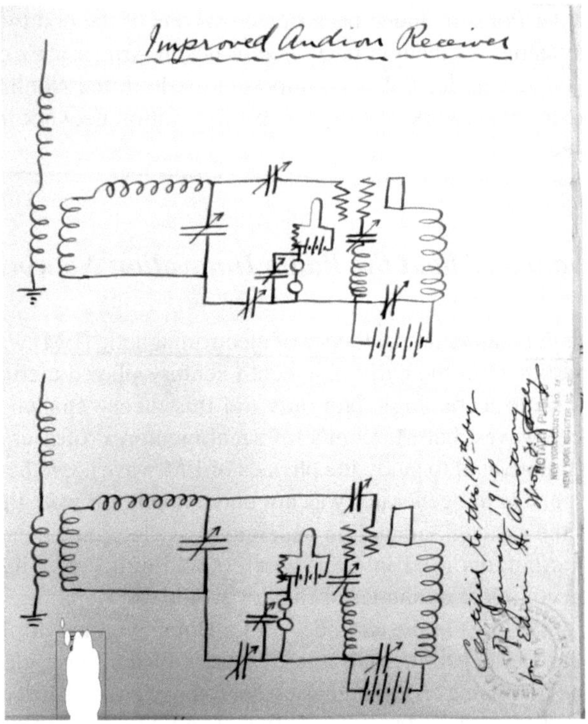

Fig. 5.7 Original sketch of improved audion circuit signed by Edwin Armstrong for wireless telegraphy. Courtesy Columbia University Archives

cation services. Marconi at first tried to capture some of the overseas market that was then being sent by underwater cable. But he quickly demonstrated a new application in ship-to-ship and ship-to-shore communication with his wireless system that was helped in part by the tragic demise of the Titanic. Just as the airplane evolution followed on the success of the internal combustion engine, so radio emerged from the success of the telegraph and the wireless inventions.

The radio network shows the importance of the role played by university nodes. For example, Columbia University and its electronics laboratory were one of the major contributors in that Edwin Armstrong was trained and continued to work there. We have to also acknowledge that the Marconi node was more that Guglielmo Marconi himself. The Marconi Company in both Britain and the US employed many engineers and technicians. Later RCA played a similar role headed by David Sarnoff who had at one time worked for Marconi. Also amateur wireless and radio organizations brought in many more players who developed and tested new technology in radio electronics and communication.

Another institution that has links to other innovation networks is the Technical University Berlin (Technischse Hochschule Charlottenburg) This institution is not the same as the University of Berlin. In the upper left hand corner of the wireless/radio network of Fig. 5.4 one can see the name of Adolf K.H. Slaby and Count George van Arco who both patented a system of wireless telegraphy similar to Marconi's early wireless projects circa 1901. Slaby knew the Chief Engineer of British Post Office, William Preece and worked with Marconi in early wireless experiments. He also had a personal link to Siemens, a degree from 'Berlin Tech', and studied with Professor Franz Reuleaux in Mechanical Engineering. Reuleaux is known as the *'father of kinematics of machines* (Chap. 2). Reuleaux was a consultant to Otto in the internal combustion network (Chap. 3) and was also an early supporter of Lilienthal in his education at Berlin Tech. (Chap. 4). That Reuleaux is linked to several innovation networks is not unusual. There were many key nodes who themselves had a wide network such as Siemens, Edison and Bell. These nodes were often not the principal inventors but served as 'connectors' such as Octave Chanute in the aviation network.

Marconi and Armstrong

It is interesting to note that although the Marconi company was a pioneer in digital wireless telegraphy, it was slow to move into analog radio transmission. An interesting communication showing Marconi's tentative interest in Armstrong's analog radio circuitry is a telegram from G Marconi to F.M. Sammis of the New York Marconi Company office. Frederick Sammis the chief engineer had telegraphed the news of Armstrong's new feedback circuit to his chief.

Marconi House, Strand
London, 22 January, 1914.

F.M. Sammis, Esq
Marconi Co., etc.
New York
Dear Sir

I am much interested in the information which you give in your letter dated January the $13th$ addressed to the Company, in regard to the results obtained at Professor Papin's laboratory at Columbia University NY. As we have carried out tests which have given extraordinarily good results in connection to the modifications to the circuits of the so-called audion and other vacuum detectors it will be necessary for us to know exactly what the device of Mr. Armstrong is, and whether it has been patented, and if so when.

We have patents dating back six months on arrangements which may be similar or preferable to those shown to you at Columbia. – – – – –

Yours very truly, G. Marconi

This telegram is in the Armstrong Archive of Columbia University. Sammis is Frederick Sammis the chief engineer of American Marconi. According to Lewis (1991) Sarnoff had met Armstrong and was given a demonstration of Columbia's new highly sensitive regenerative circuit. This Sarnoff reported to Sammis who then cabled Marconi. Sarnoff had reported that: *"I would state that the results obtained with Mr. Armstrong's receiver are sufficiently convincing to warrant our most careful investigation of his patents and circuits etc. For I believe his device had tremendous advantages–."*

The telegram above shows that the Marconi firm in New York, through Sarnoff and Sammis, had its so-called ear to the ground regarding new advances in electronic circuits. It also seems to show that Guglielmo Marconi himself took a skeptical view of technology not invented at Marconi Co. Thus the NIH principle of 'not invented here', often attributed to modern corporations such as IBM, was flourishing a century ago.

5.4 Wireless and Radio Books and Magazines

Finally the spread of the knowledge of wireless and radio was accelerated by the publication of new books and magazines even as the technology was being transformed. The Italian book by Mazzotto (1906) (Fig. 5.8) appeared even before vacuum tube circuits would revolutionize the technology in the next decade. While we sometimes credit our own era with unique innovation capabilities, such as the distribution of

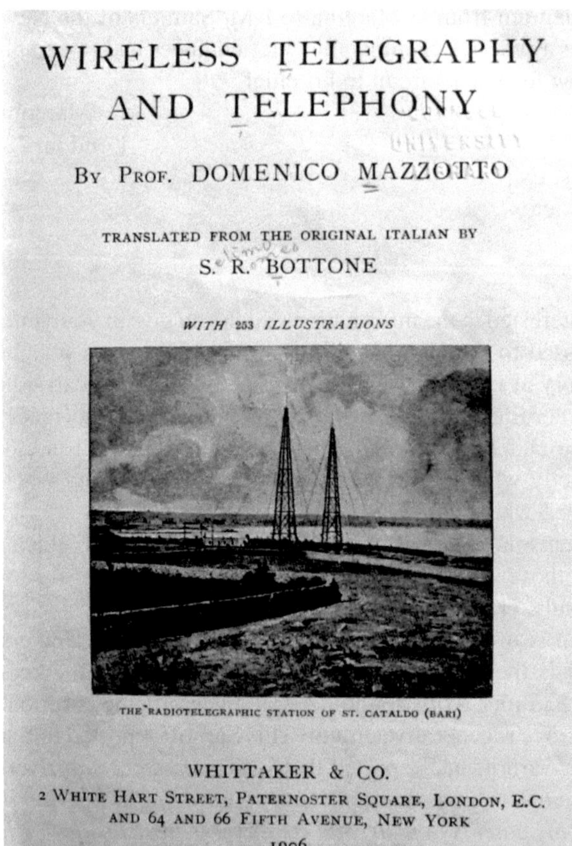
WIRELESS TELEGRAPHY
AND TELEPHONY

BY PROF. DOMENICO MAZZOTTO

TRANSLATED FROM THE ORIGINAL ITALIAN BY
S. R. BOTTONE

WITH 253 ILLUSTRATIONS

THE RADIOTELEGRAPHIC STATION OF ST. CATALDO (BARI)

WHITTAKER & CO.
2 WHITE HART STREET, PATERNOSTER SQUARE, LONDON, E.C.
AND 64 AND 66 FIFTH AVENUE, NEW YORK
1906

Fig. 5.8 One of many early books on wireless telegraphy, by Mazzotto (1906)

knowledge though the World-Wide Web, earlier ages of invention had different communication channels and social networks that also resulted in exponential avalanches of knowledge and technical breakthroughs.

An early book is L.J. Fahie's *A History of Wireless Telegraphy,* 1901 second edition. Fahie was a member of the British IEE or Institution of Electrical Engineers. Even the great British physicist J.A. Fleming wrote the book *The Principles of Electrical Wave Telegraphy, 1906.*

Another book published in German by Ruhmer in 1908, was *Wireless Telephony* and quickly translated into English such was the interest in wireless technology. Maver's *Wireless Telegraphy and Telephony,* appeared in 1910.

An American work is A. Frederick Collins's *Wireless Telegraphy; Its History, Theory and Practice,* published in 1905, less than five years after Marconi had set up his company in the UK. Collins wrote about Hertz's experiments on electromagnetic waves, EM wave transmitters and detectors including coherers of which Collins had invented one of his own.

Another popular book was A.P. Morgan's *Wireless Telegraphy and Telephony Simply Explained,* 1913; "A Practical Treatise; Very Fully Illustrated" so the inside cover proclaims. Morgan dedicated this book to Nicoli Telsa.

Thus the beginning of the technical innovation avalanche, that begun around the 1890s generated the means of sustaining and expanding the exponential rush to create ever new variations of the innovation similar to the 'dot.com' revolution of the 1990s a century later.

On the magazine front there were long running ninenteenth century magazines such *Scientific American* in the US and *Engineering* in the UK. A partial list of English language magazines publishing articles about wireless and radio includes;

Scientific American

Electrical World [1906, NY]

The Electrical Engineer [A serious journal published by the Institution of Electrical Engineers, London]

The Electrician [London]

The Wireless World [1913, Published by Marconi Co., London]

The Wireless Age

The Electrical Experimenter[1916]

QST [Published by the American Radio Relay League, 1915.]

Wireless Engineer [1923, London]

For example the November 23, 1907 issue of the *Scientific American,* carried a full page story about Marconi's Ireland transmitter at Clifden Station, describing with photographs, the "Marconi Wireless Telegraph System". This station was destroyed by a fire during the Irish rebellion in 1922. The station was transferred to Wales.

In *Electrical World,* of 1913 appeared an announcement of the new American organization, The Institute of Radio Engineers, or IRE.

The August 1916 issue of *The Electrical Experimenter,* had an article "The Vacuum Detector: How it works?" This article was reviewed by J.H. Morecroft and E.H. Armstrong of Columbia University, as well as Lee de Forest. Although de Forest had invented the triode vacuum tube in 1906, he did not fully understand how it worked until the careful experiments of Armstrong and others at Columbia University circa 1914. Thus radio amateur enthusiasts were able to get this knowledge very shortly after it was discovered, not unlike the present age of Google and the World Wide Web.

As an example of the maturity of the communications systems at the dawn of the twentieth century we describe a snapshot of the public announcements related to wireless telegraphy published in the English journal *The Electrician: a Weekly Illustrated Journal of Electrical Engineering, Industry and Science*, from October 1901 to April 1902. The principal focus of this journal was on mature electrical industries such as electric power, electric motors, transportation and wire-born magnetic telegraphy. One can find articles concerning patents, new companies and professional society meetings. There were also some tutorial articles such as lectures by Oliver Heaviside on electromagnetics. But by and large this journal was boring reading for the lay person, except for the new announcements relating to wireless that promised to challenge the conventional telegraph land and cable business.

The Electrician 1901–1902 Volume XLVIII, London

Greater than 80 articles in the Index under "Wireless Telegraphy"

Issue Oct 25, 1901

Marconi's wireless system used to report yacht races off New York harbor;
Reported in the New York based *Electrical World.*

Issue Dec 20, 1901 [p. 329]

"The current week opened with the startling announcement that Mr. Marconi had succeeded in transmitting wireless signals across the Atlantic." St John's Newfoundland received 'waves' transmitted from Cornwall, UK- the Morse letter 'S' at intervals of five minutes as prearranged. "a date to be remembered in the history of wireless telegraphy".

Issue Jan 17, 1902 [p. 500]

Editorial comment raises skepticism about Marconi's transatlantic transmission.

Issue Feb 14, 1902 [p.657]

Published tutorial on Maxwell's equations of electromagnetism by Oliver Heaviside using vector calculus.

Issue Feb 28, 1902 [p.721]

"Wireless Telegraph Berlin" Article reports on competitive trials in Germany between a wireless system of Count Arco and Prof Slaby (Allgemeine Elektricitäts Gesellschaft) and Prof Braun (Siemens and Halske) as far as 200Km.

[p.730] Abstract from *Le Figaro*, "The New Branly Radio-Conductor" reports a new bi-metallic coherer by Mr. Edouard Branly of Societe Francaise des Telegraphes Sans Fil.

[p. 732] Editorial reply to Mr. Marconi's claim that the English electrical press was "being bowed to the exclusive support of cable interests."

Issue March 7, 1902 [p. 761–762]

A report that Dr. J.A. Fleming , FRS, Professor of Electrical Engineering in University College London has rendered considerable assistance to the Marconi Wireless Telegraph Company.

Issue April 4, 1902 [p. 926–927]

Reprint of a letter of Dr. Slaby of Germany in the *Times {of London}* stating that Slaby has not used ideas from Marconi's 1897 wireless tests at which Slaby was present. "The Slaby-Arco system is not based on those earlier tests."

Issue of April 4, 1902 [p. 943]

'Recent Wireless Telegraphy Developments"

reports rivalry between the Marconi and Slaby-Arco systems. Need for international understanding. Seimens-Halske system of Braun will begin litigious warfare against the Marconi system in England and the Slaby-Arco system in Germany.

[p. 947] Reports on Canadian government agreement with Mr. Marconi.

[p. 952] "Mr. Marconi's Latest Achievement in Ocean Wireless Telegraphy" Reports transmission from the ship Philadelphia to station in Cornwall England a distance of 1551 miles.

Issue April 11, 1902 [p. 967]

Further statements between Professor Slaby and Mr. Marconi.

Issue April 18, 1902

Discussion on "German Views on Wireless Telegraph" – another description of legal claims of the Marconi, Slaby-Arco and Braun systems.

[p. 1006] "Wireless Telegraphy in Spain" Three page article with circuit, map and photos of a wireless system of Senor Julio Cervera Baviera in which he transmitted signals between Spain and Morocco across the Strait of Gibraltar.

[p. 1037] "American Wireless Telephone and Telegraph Co." Article states the claim that "The company owns what is described as the basic patent No. 350299 dated Oct 5 1886 granted to Prof, H.S. Dolbear, 'the original discoverer of wireless telegraphy'" "No person or company with any system of communication without wires can operate in the United States – without obtaining a license from the owners of this patent."

The pace of announcements in wireless telegraphy in the six months of *The Electrician,* October 1901 to April 1902, was almost breathtaking, beginning with the ship to shore tests of Marconi off the shore of New York City to the competing systems of Slaby-Arco and Braun in Germany. We also read of Marconi's claim of transatlantic transmission and the skepticism showed by the cable community.

Of particular interest is the name of the Spaniard Julio Cervera who conducted tests on a working wireless system and is never mentioned in the standard histories. We also read of 'prior art' claims of the American company announcing intentions to sue for license infringement. Also in these pages we have direct evidence for historical network links between Marconi, Fleming, Branly, Slaby, Arco, Braun and several corporate institutions in England, Germany and France.

We also can see the exchange of communication between different papers such as the *Times of London, Le Figaro, Electrical World* in New York, each quoting one another. In the age of the Internet we sometimes boast of exchange of information at the speed of light, but we forget that circa 1900, information also traveled at the speed of light through the thousands of miles of land and ocean cables around the world in what (Standage 1998) called the "Victorian Internet".

5.5 Nicola Tesla and the Claim: 'Father of Radio'

To some believers in the hero-inventor theory of innovation, the Serbian-American Nicola Tesla is the father of wireless and radio. What is the origin of this myth? In the 1890's he had proposed a wireless system to transmit and receive communications and power. He obtained some patents for these ideas. However he did not build a full scale working system as did Marconi and Slaby-Arco or Braun. His real focus and contributions were to generating power, not transmitting information. Tesla did inspire some in the wireless network such as Lee de Forest. Tesla belonged to the electric power generation innovation network not the wireless network.

As part of his work in electric power generation, he designed electrical transformers (sometimes called a Tesla coil) that would boost voltages high enough to create arcs in the air or in partial vacuums. Spark-gap technology for generating electromagnetic waves for wireless transmission was an early component of wireless systems but was replaced in a short time by high frequency vacuum tube transmitters. His work was not unique in this endeavor as others such as Stone and Lodge, were both working before and after Tesla on spark gap E-M wave generation as well as transformer design.

Tesla received his education under the Austro-Hungarian empire. After schooling he worked for a telegraph firm in Hungary. He migrated to France and took a job with Edison-France. He then sailed for New York and took a job with the Edison Company in New York who was building electric generation systems for electric light systems. Tesla proposed to improve Edison engineers' designs for generators and was rebuffed. He eventually allied himself with Westinghouse Electrical Manufacturing Co. in Pittsburgh and the movement to replace the Edison direct current generation with alternating or AC current systems. His most important contribution was the concept of a rotating magnetic field motor-generator. Tesla was recognized for this accomplishment with the naming of the international unit for magnetic field density or 'Tesla'. MRI or 'Magnetic Resonance Imaging' systems today employ magnetic fields on the order of 1 'Telsa' that is 20,000 times the value of the magnetic field at the surface of the Earth.

The source of the appellation "Tesla as the inventor of radio" is the 1943 US Supreme Court decision in a lawsuit filed by the Marconi Wireless Co against claims of patent infringement of the US government stemming back to WWI. After several earlier Appeals Court decisions, the Supreme Court ruled that the Marconi claims were not valid because the operating patents circa 1901, were superseded by patents of Tesla, Lodge and Stone. Zealots of Tesla used this to say that the Supreme Court had anointed Tesla as the inventor of radio. However in an analysis of the Court decision by Prof A. David Wunsch of U Mass Lowell he says: "thus while the Supreme Court declared the Marconi patent invalid, it affirmed prior patents by not only Tesla, but by Lodge and Stone as well" (Wunsch 1998) "The Supreme Court never determined that Tesla invented radio." The patents in question had to do with resonant circuits for the transmission and receiving electromagnetic waves.

There were other important components to wireless and radio such as the coherer rectifier and later the vacuum tubes of Fleming and de Forest as well as the super-heterodyne circuits of Armstrong circa 1918. Unfortunately media outlets including the *New York Times* as well as PBS in a special show on Tesla have interpreted this legal decision as an imprimatur of "Father of Radio" status for Tesla. But legal decisions rarely get the technical history correct, and most historians of technology agree that wireless and radio was a collaborative effort that evolved over a period of some 30 years from the time of Hertz's experiments to the radio electronics of Armstrong circa 1920.

That Tesla was a clever engineer who created new designs for electrical technology is not debated. However he was not an isolated inventor. He was enmeshed in a rising tide of electric power generation around the world as well in the telegraph communication industry that had matured since the 1850s and 1860s with the advance of ocean cable technology. He was not leading this exponential growth in innovation, it was carrying him. (see Fig. 5.3) Tesla was part of a social network of engineers, scientists, artisans, industries and entrepreneurs and professional societies all moving the frontiers of electrical power, lighting and communication. The interrelationships between the magnetic telegraph, electric generator, electric lighting and wireless networks is shown schematically in Fig. 5.9.

For some the Tesla myths became part of a Tesla cult, partly because in his later years he became somewhat of a futurist, proposing fantastic technologies for the future that some identify with our current world. But envisioning a future world is not the same as creating one.

5.6 Interlinked Innovation Networks

Using the network idea we can see how it took over four decades for engineers to understand the vacuum tube and the circuits that enabled the radio of the 1920s and 1930s. This evolution began with the Edison tube and Fleming's diode in the 1890s. Then we have Lee de Forest and the triode vacuum tube of 1906 and a decade later Armstrong's regenerative and superheterodyne circuits of 1917. And finally van der Pol is able to create a mathematical model in the late 1920s from which other technical ideas emerged such as an electronic model of the human heart and later the field of bioelectronics and the electronic pacemaker.

If one were to construct an innovation network for the telegraph, electrical generation, electrical lighting, wireless telegraphy and radio, we would expect there to be connections between these structures producing a super-electrical technology network as imagined in Fig. 5.9. However the medium of print technology makes this

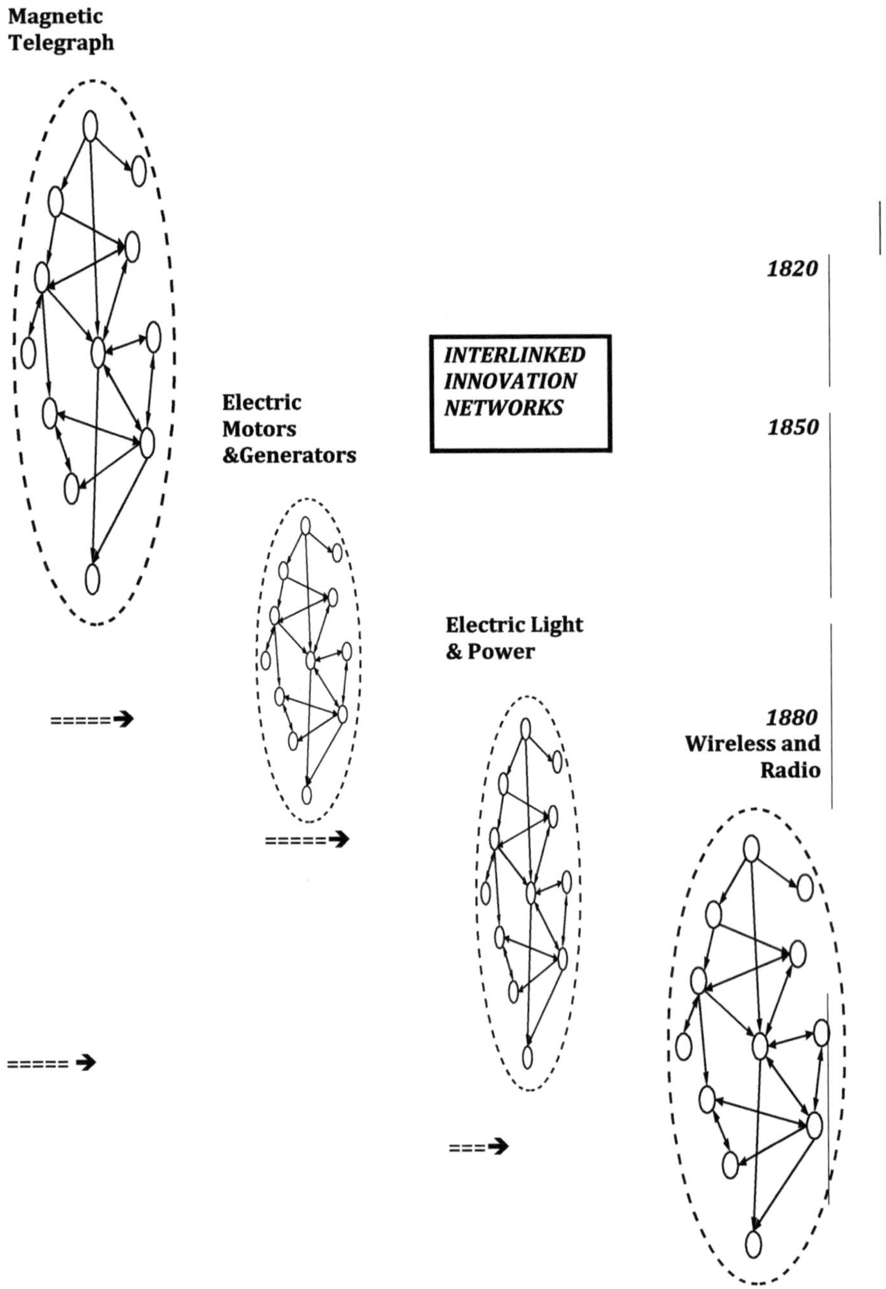

Fig. 5.9 Schematic sketch of interlocking telegraph, electrical power and light, and wireless networks

super network difficult to display. We trust that other authors will construct one for the web that one can explore more easily.

5.7 Ørsted and the Early Electromagnetic Theory Network

We end this chapter by looking back at the roots of electromagnetic theory that led to the wave nature of electrical phenomena of Maxwell and Hertz. It is easy to understand the exponential rise in wireless network nodes at the end of the nineteenth century since there was already in place a worldwide telegraph communication network and a growing telephone network. But what were the links between electrical scientists a century earlier? The work of L. Pearce Williams on the *Origins of Field Theory,* (Williams 1965) outlines the principal players in electrical science from Gilbert (1600) to Maxwell (1864). Two key scientists were Benjamin Franklin [1706–1790] and Hans Christian Ørsted [1777–1851]. Franklin had studied the properties of electricity and had discovered the connection between lightning and electricity. Ørsted discovered the connection between electrical current and magnetism.

Franklin in his autobiographic sketch, *The Life and Essays of the Late Doctor Benjamin Franklin: Written by Himself,* (Franklin 1814) described how he became interested in electricity. He begins by reciting a litany of early discoveries in magnetism and electricity from the time of Gilbert (1600), Boyle, Otto Guericke and several scientists in England, France and Germany from 1706 to 1742. Then he writes,

"Collinson, about the year 1745, sent to the library company of Philadelphia an account of these experiments— Franklin with some of his friends immediately engaged in a course of experiments; the result of which is well known.—His [Franklin] observations he communicated in a series of letters to his friend Collinson: the first of which is dated March 28, 1747."

Peter Collinson was a lifelong friend who obtained books for Franklin for the Philadelphia library (Isaacson 2004). Though Franklin was born in the new world colony of Boston, Massachusetts, he was in cultural contact with the scientific advances in Europe. In 1724 as a young man, he traveled to London for a year. He was able to publish his findings in London in 1751. Later in his life, after his significant electrical researches, Franklin traveled again to England and France. His name appears in the balloon network in Chap. 4, because he witnessed the early assents of the Montgolfier brothers in Paris.

Ørsted grew up in Denmark and studied at the University of Copenhagen, studying pharmacology, mathematics, physics and chemistry. Perusal of his scientific works reveals his contact with and/or knowledge of many other scientists in the early nineteenth century such as Ampere, Arago, Chladni, Davy, Guy-Lussac, Faraday, Humboldt, Lavoisier, Seebeck, and Volta, He also communicated with scientific institutions such as the Academy of Berlin, and the Royal Society of London. Most of his correspondence took place before the early telegraph systems circa 1840.

Another luminary of electrical science Michael Faraday [1781–1867], in the title page of his work on "experimental researches in electricity", (1855), lists his affiliations as "Foreign Associate of the Academy of Sciences of Paris, Petersburgh, Florence. Copenhagen, Berlin, Gottingen, Stockholm, Munich, Bruxelles, Vienna, Bologna Etc. Etc. [sic]" (Williams 1971).

Faraday's American counterpart Joseph Henry had a different set of influences. Born in a small town north of New York City, he was interested as a teenager in the theatre when he accidently discovered a 1808 physics school book by George Gregory, *Lectures on Experimental Philosophy, Astronomy and Chemistry,* published in London. While teaching at the Albany Academy from 1827–1832, Henry made remarkable discoveries relating to the building of large electromagnets and self induction of electric current. These discoveries led to the magnetic telegraph, the creation of arc generators used in early wireless systems and the inductor circuit element used in electronic circuits to generate and detect electrical signals. But while Faraday could boast connections to all the top scientific academies of Europe, Henry had only *Silliman's Journal* [The American Journal of Science] and the *Journal of the Philosophical Society of Philadelphia* in which to publish. However Henry did seem to be aware of the research of Ørsted and Ampere during his time at Albany. After his move to Princeton University, in 1832, he visited London in 1837 and came into contact with some European scientists. During a later European trip he met Ampere and they did some experiments together. Despite the War of American Independence and the War of 1812 with England, the scientific bonds between the old and new world were strong, leading to important research results from Joseph Henry.

The study of the correspondence and links between scientists and mathematicians of the sixteenth to nineteenth centuries and the various academies in different countries of Europe would itself make a very interesting topic but is beyond the scope of this monograph. Instead we move into the twentieth century and explore the links between the radio electronics network and the mathematical theory known as chaotic dynamics and fractals in Chap. 6.

References

Collins AF (1905) Wireless telegraphy: its history, theory and practice. McGraw Publication, New York

Douglas SJ (1987) Inventing American broadcasting 1899–1922. The Johns Hopkins University Press, Baltimore

Fahie JJ (1901) A history of wireless telegraphy, 2nd edn, revised. W. Blackwood and Sons, Edinburgh

Faraday M (1855) Experimental researches in electricity. Richard Taylor and William Francis, London

Fleming JA (1906) The principles of electrical wave telegraphy. Longmans, Green, London

Franklin B (1814) The Life and essays of the late doctor Benjamin Franklin. William Fessenden, Brattleborough Written by himself

Gregory G (1808) Lectures in experimental philosophy. Astronomy and Chemistry, London

Isaacson W (2004) Benjamin Franklin: an American life. Simon and Schuster Paperbacks, New York
Larson E (2006) Thunderstruck. Crown Publication, New York
Lewis T (1991) Empire of the Air: The Men Who Made Radio. E. Burlingame Books, Harper Collins, New York
Maver W (1910) Wireless telegraphy and telephony: a handbook of. Maver's Publication, New York
Mazzotto D (1906) Wireless telegraphy and telephony. Whittaker and company, New York
Morgan AP (1913) Wireless telegraphy and telephony simply explained. Norman Henley Publication, New York
Orsted HC (1998) Selected scientific papers of Hans Christian Orsted. Jelved K et al. (ed). Princeton University Press, Boston
Ruhmer E (1908) Wireless telephony: in theory and practice. Crosby Lockwood and Son, London
Silverman K (2003) Lightning man: the acursed life of Samuel B. Morse. da Capo Press, Cambridge
Standage T (1998) The victorian internet. Walker Publishing Company, New York
Wikipedia subject "Invention of Radio" and "Timeline of Radio"
Williams LP (1965) The origins of field theory. Random House, New York
Williams LP (1971) Michael Faraday, a Biography. Simon and Schuster, New York
Wunsch AD (1988) Misreading the Supreme Court: a puzzling chapter in the history of radio, internet, antenna, newsletter of the mercurians, in the Society for the history of technology, Edwin armstrong papers, Columbia University Archives, New York vol 11(1), November 1988

Chapter 6
Social Networks in Nonlinear Dynamics, Chaos and Fractals

Abstract We show how the seeds of twentieth century mathematical Chaos Theory grew in the networks associated with the internal combustion engine and radio electronics. We trace the mathematical discoveries of Henri Poincare in France at the beginning of the twentieth century to the fluid mechanics weather models of Edward Lorenz at MIT and electrical circuits of Yoshisuke Ueda in Kyoto a half-century later. This example was chosen because of the author's close familiarity with chaotic and nonlinear dynamics. This example and the radio network show that the ideas of a network community for innovation go beyond the examples from mechanical science in Chaps. 1–4.

6.1 Introduction to Modern Chaos Theory

In the late 1970's, the concept of unpredictability in deterministic science, what we now call *Chaos Theory*, became a major research area in nonlinear science. By the late 1980's thousands of research papers had been written, popular books such as James Gleick's *Chaos* 1987 were published and the American public TV series *NOVA* (1989) screened a popular show on "The Strange New Science of Chaos" in which this author appeared. How did the mathematical field of 'nonlinear science' emerge from academic esoterica in the 1970's to become a cocktail circuit topic of discussion by the early 1990's?

One of the phenomena that resulted from this popularization was the explosion of the use of the word *chaos* in the world's major newspaper headlines beginning around 1984 (Fig. 6.1). In 1980 there were recorded approximately 50 headlines a year using the word 'chaos' in major newspapers, whereas in the year 2000 there were more than 3000 headlines using the term (Source; LEXIS NEXIS). In 1998, the *New York Times* proclaimed "Congress in Chaos" and the same year the *Times* of London posted a headline, "Currency Chaos Spills Over into Bond Market". As we

F. C. Moon, *Social Networks in the History of Innovation and Invention*, History of Mechanism and Machine Science 22, DOI: 10.1007/978-94-007-7528-2_6,
© Springer Science+Business Media Dordrecht 2014

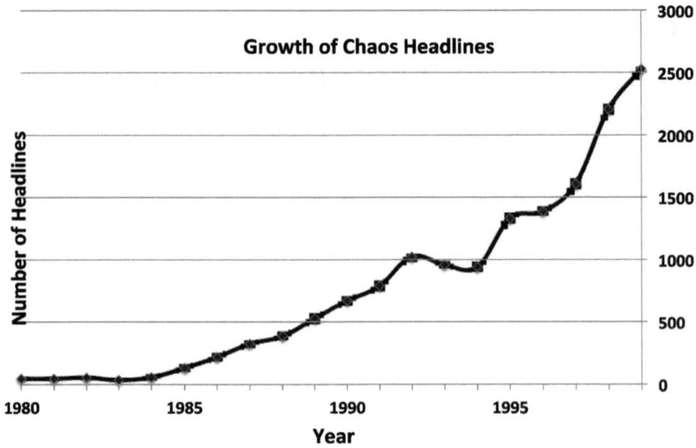

Fig. 6.1 Graph of the number of headlines per year using the word '*chaos*' in the world's newspapers (*Source* LEXIS NEXUS)

shall see below, the rise of the popular interest in chaos lagged more than a decade behind the rise of interest in the mathematics, science and engineering communities.

This chapter may be difficult to follow for the layperson not familiar with applied mathematics. It's not like IC engines where most people have some contact with automotive components. The field of nonlinear science also has a number of subnetworks of which we shall describe five. Most of the names in the nodes of these networks are only known to the aficionados of chaos theory. Although this field appears outside the realm of technology, it grew naturally out of the networks of two other technologies as well as abstract mathematical analysis.

For the layperson, one way to view these networks is to read it as a map of a foreign country. We trace the roads and highways past names of villages we can't pronounce searching for a route to our destination. The nodes in the networks for chaos theory are like those foreign towns and the route is the series of links that connect our understanding of how technology and science are closely connected. Two of the three main sources of chaos theory had origins in engine technology (Chap.3) and radio electronics technology (Chap.5). The evolution of chaos theory is an example of overlapping networks and the interconnectedness of science and technology.

We also hope to demonstrate how history is shaped by those who write about it. When mathematicians were asked to relate the history of nonlinear science and chaos theory, they described the principal nodes and links in terms of mathematicians like themselves in what we will call "The Standard Model". Physicists who later attempted to define the historical origins of nonlinear science added a few physicists to the 'standard model'. Both models ignored or marginalized the role of engineering technologies in providing experimental and mathematical models from which the phenomena of unpredictable dynamic behavior became a problem to study by mathematicians in the standard model.

Names of the principal nodes in the 'standard model' of nonlinear dynamical systems (See e.g. Guckenheimer and Holmes 1983):

Henri Poincare [1854–1912]: French mathematical physicist. The first to recognize the unpredictability of Newton's equations of motion for the planets.
George Birkhoff [1884–1944]: American mathematician who followed up on Poincare's conjectures. Laid the math foundation for chaos theory.
John E. Littlewood [1885–1977]: British mathematician who worked with Cartwright on the van der Pol circuit equations.
Mary Cartwright [1900–1998]: British mathematician. Proved some important properties of the nonlinear circuit equation of van der Pol.
Solomon Lefshitz [1884–1972]: Mathematician, Princeton University, One of the pioneers of topology. Helped introduce Russian mathematicians to the US.
Steven Smale [1930]: Mathematician, University of California, Berkeley. Developed important mathematical concepts of chaotic dynamical systems.
Edward Lorenz [1917–2008]: MIT meteorologist. Early pioneer in chaos theory of weather models.
Benoit Mandelbrot [1924–2010]: Polish-French mathematician, worked at IBM, Yale. Known for his fractal theory. Co-winner of the Japan Prize, 2003.
James A.Yorke [1941]: Mathematician University of Maryland. Introduced the term 'chaos' in 1975 in a paper with a student, Li. Co-winner of the Japan Prize 2003.
Mitchel Feigenbaum (1944): Mathematical physicist, Rockefeller University, NY. Developed mathematical tools to study chaos. Winner of Wolf Prize in physics.

Chaos theory had its origins in the Newtonian dynamics of celestial mechanics. Chaos theory was also motivated by problems in fluid mechanics (meteorology and aviation), radio electronics and the dynamics of internal combustion engines and vehicles. By the 1920's and 1930's, unpredictable behavior in predicting weather, the dynamics of aircraft, noise and electrical static in radio technology and unwanted vibrations in internal combustion engines and electrical generating systems came to the attention of theoretical engineers and eventually to that of applied mathematicians. Two of the principal mathematical models of chaos theory were the *Duffing* equation, and *van der Pol* equation, each posited by engineering scientists, who were employed in industry, namely George Duffing (Deutz Motors, Cologne) and Balthazar van der Pol (Phillips Radio Corp, Netherlands). A now classic example of chaotic vibrations in machines is gear rattling, illustrated in Fig. 6.2. Small wear or play in the gears creates a gap in the contact, that when closed produces impact, leading to noise in the machine employing these gears. Similar examples occur in electronic systems.

Poincare and the Early Nonlinear Mathematicians

The origins of modern chaos theory stem from the laws of dynamics posited by Isaac Newton (c.1686) whose mathematical development reached its peak in the

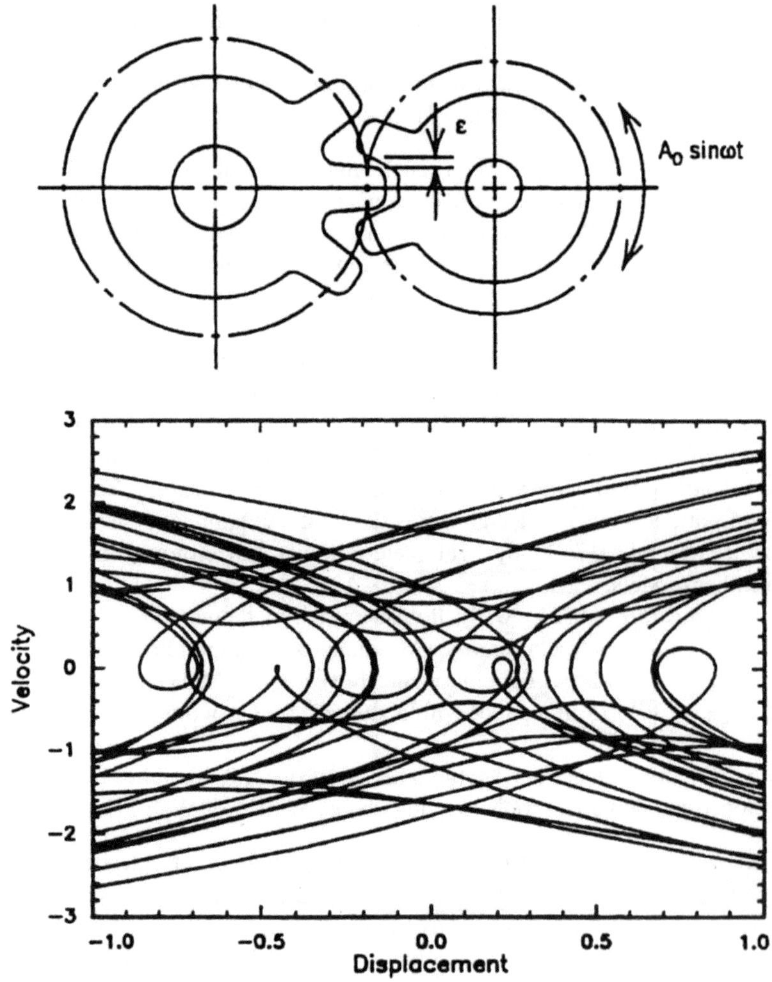

Fig. 6.2 *Top* Example of machine chaos in rattling gears found in automobile transmissions. *Bottom* Computer simulation of machine chaos in gears plotting gear movement versus gear velocity

nineteenth century. These laws were framed in the language of differential calculus. The mathematical logic of the time supposed that if the positions and velocities of all the mass particles in the universe were known at some initial time, then the motions of the universe would be predictable forever into the future; hence the phrase "*a clockwork universe*" when referring to Newtonian physics. Near the turn of the twentieth century, however, the mathematical physicist Henri Poincaré [1854–1912], examined this premise of predictability in his study of the motions of the planets for which he won a prize from the King of Sweden in 1890. Poincare wrote this prescient phrase in chaos theory; (See , Holmes 1990) "*it may happen that small differences in the initial conditions produce great ones in the final phenomena*"— (Poincare' 1908).

Poincare' uncovered a remarkable truth; that predictability required an infinite precision of knowledge about the initial positions and masses. Any small uncertainty in the initial conditions could lead to a large divergence of motions in the future. That is, *predictability was impossible*. And this was two decades before the formal development of quantum mechanics and the discovery of its famous *uncertainty principle*. What Poincaré had uncovered was an uncertainty principle for Newtonian mechanics.

Mathematicians who followed Poincare in exploring the message of unpredictability in physics, included George D. Birkhoff (Harvard Professor) in the United States and A. A. Andronov in the Soviet Union. Also Dame Mary Cartwright and John L. Littlewood from Cambridge were working on the mathematics of radio dynamics during World War II in England. Birkhoff called this new field *nonlinear dynamical systems*, and some mathematicians prefer this term to *Chaos Theory*. In the post war period Steven Smale in 1960 provided a mathematical construct called the ' horseshoe map' that would become the foundation of understanding the strange dynamics of chaos a dozen years later. There are dozens of popular books on chaos theory. For the interested reader, the most published one is James Gleick's Chaos: Making of a New Science, 1987, still in print today.

Though mathematicians claim that chaos theory had it roots in mathematical analysis, there is evidence that emerging technologies of the late nineteenth and early twentieth century also spawned some of the seeds. For example, in the speed control of rotating machinery it was known that machines could experience unsteady 'hunting' vibrations. The problem of hunting dynamics in steam engines was studied by James Clerk Maxwell, who was also one of the nodes in the electronics network. Another source of nonlinear and chaotic dynamics was electronics itself, especially the new vacuum tube circuits. A Dutch electronics engineer, Balthazar van der Pol, and colleague J. van der Mark in 1927, are recognized as the first to document unsteady vibrations in electronic circuits using vacuum tubes. The mathematical equation that models this behavior in called today the *van der Pol* equation.

In the arena of aircraft dynamics, as the development of internal combustion engines allowed aircraft to fly at higher speeds, pilots noticed dangerous vibrations of the wing and fuselage that sometimes tore the wings off the plane. Another Dutch engineer Den Hartog wrote a book in the 1930's that outlined mathematical models for these fluid-structure vibrations called *aeroelastic flutter*. A similar phenomenom occurs in electrical transmission lines that are observed to vibrate violently in strong winds. In the former Soviet Union of the 1930's (today's Russia) three mathematicians Andronov et al. (1966) wrote about a collection of technical nonlinear dynamics problems involving both mechanical and electrical technologies.

6.1.1 Edward Lorenz and Weather Chaos

Everyone experiences the unpredictability of weather. The science of weather is governed by the field of fluid mechanics that is another source of modern chaos theory. Its roots go back to Leonardo da Vinci (Chap.3) and his observation of turbulent flows, later studied by Lord Rayleigh in the nineteenth century. The avoidance of turbulence in the flow of air over aircraft wings was a key element in the design of aircraft. Fluid mechanics is also at the heart of understanding weather and meteorology that also evolved with the development of long distance ballooning and aviation.

One of the major breakthroughs in chaos theory came in 1963 from a MIT meteorologist Edward Lorenz, who examined the predictability of the fluid mechanics associated with the Earth's weather. In a 1993 memoir, Lorenz says he was guided by Birkhoff's earlier work on dynamical systems after attending a few lectures of Birkhoff at Harvard. Using a new generation of electronic computers, made faster by the discovery of the transistor in 1957, Lorenz showed that even simple mathematical models of weather lead to complex unsteady patterns of temperature and velocities. Despite the rapidly advancing progress of computers, long-term weather prediction was impossible. Although these ideas were published in a reputable scientific journal, they were all but ignored for a dozen years except for a handful of applied mathematical scientists. Lorenz's paper was not published in a mathematics or regular physics journal but in the *Journal of the Atmospheric Sciences*, March 1963. In the abstract of this paper he writes of the mathematical model for the fluid convection of the atmosphere:

"*All the solutions are found to be unstable and almost all of them are nonperiodic. The feasibility of very-long-range weather prediction is examined in the light of these results*".

Before Lorenz's death in 2008, he wrote a lay-person's book about his research into the predictability of weather called *The Essence of Chaos*1993. He describes earlier laboratory experiments of R. Hide in 1953 on fluids that rotate in an experiment simulating the Earth's atmosphere in which he observed instabilities of these motions. Lorenz also describes the history of computational models to predict the weather dating back to World War, to recent programs with powerful computers using five million variables. Of these attempts he wrote;

"*Almost without exception, the models have indicated that small differences (in data) will amplify until they are no longer small—in spite of the good forecasts that they (the models) produce at short range, we are more or less forced to conclude that the atmosphere itself is chaotic*".

Before Lorenz, there were other scientists, engineers and mathematicians who decades earlier had observed strange unsteady motions or noise in their experiments or numerical calculations. In 1961, electrical engineering student Yoshisuke Ueda at Kyoto University, discovered strange oscillations in computer simulations of electrical circuits. Earlier Van der Pol at Phillips Research had reported "irregular noise" in his experiments with an electronic oscillator in the magazine *Nature* in 1927. So what was new about the dynamics of chaos discovered in the 1970's and 1980's?

What was new about Chaos was the discovery of an underlying order in these unpredictable motions called *bifurcations and fractals*. This underlying order was common to all kinds of systems from animal and plant ecology to meteorological dynamics and electrical circuits. Lorenz was studying mathematical models of weather and Ueda was modeling the nonlinear dynamics of electrical circuits. In the 1970's, patterns of chaos in the population dynamics of animals and ecosystems were discovered by Robert May of Princeton, and strange motions of structural systems were predicted by Philip Holmes at Cornell. Lasers, transistors, ship capsize, friction devices, chick heart cells and more were discovered to have these unsteady motions and patterns of dynamic behavior in the late 1970's and early 1980's. Chaos was coming out of the woodwork it seemed, to quote the mathematician Steven Smale,

"*In the 1970's the scientific community recognized another revolution, called the theory of chaos, which deals a blow to the Newtonian picture of determinism. As a consequence, the world now knows that one must deal with unpredictability in understanding common experiences*".

Why did mathematicians use the term *chaos* and not *random* as an adjective? Random motion was understood to mean not only unpredictable, but of unknown origin. These strange dynamics were not random motions of unknown origin. They were inherent in the same deterministic equations of motion of classical physics that determine the motions of the Earth and Moon. A new name was needed and the name we now use is chaotic dynamics or *Chaos*.

6.1.2 The Early 'Geniuses' of Chaos Theory: The 1970s

Chaos theory had its share of genius innovators who magazines and writers liked to feature in the 1980's. However, as we shall see below, there was a complex network of mathematicians, scientists and engineers working toward a theory of chaos beginning in the 1960's and 1970's. The claim of authorship of the term *chaos* in nonlinear dynamics belongs to David Ruelle and Floris Takens in France in 1971 and Tien-Yien Li and James Yorke of the University of Maryland in 1975. Ruelle, a Belgian and Takens, from the Netherlands wrote a mathematical paper on the nature of fluid turbulence in *Communications in Mathematical Physics* and remarked that for some conditions "*the fluid motion becomes very complicated, irregular and chaotic*". They also used the term '*strange*' attractor to describe the fractal mathematical object that the chaotic, turbulent motion describes in an abstract universe called *phase space*.

Li and Yorke (1975) 4 years later in a paper in *American Mathematical Monthly* used the title "*Period Three Implies Chaos*". They discussed the dynamic changes in mathematical models from simple motions to chaotic motions as some parameter is varied. Thus a regular periodic motion could change to a motion with a period three times the basic period. Their theorem concluded that if such a bifurcation occurred in the dynamics of some physical system as some control parameter is varied, then it is possible that chaotic oscillations will occur at some other parameter value. Hence

Fig. 6.3 Chaos Quartet. Left to right Mitchel Feigenbaum (Rockefeller University, NY, 1986 Wolf Prize in Physics), Yoshisuke Ueda (Kyoto University), Tien-Yien Li (Michigan State University) James A. Yorke (University of Maryland, Winner of the 2003 Japan Prize with B.B. Mandelbrot) 2010 Ueda Symposium on Chaos, Kyoto Japan

the title "*Period Three-Implies Chaos*." In a recent conversation with T.-Y. Li , now a professor at Michigan State University, he remembered working as a graduate student with Jim Yorke at Maryland and finding the result at least a year earlier than was published. Li said that Yorke was in no hurry to publish it believing that not many would be interested in reading the paper. In 2003, Jim Yorke won the Japan Prize for his work on nonlinear chaotic dynamics.

One of the paths from regular motions to chaotic dynamics in many physical systems is called 'period doubling'. In regular oscillations the motion may repeat every T seconds. But as some physical parameter is changed, say the voltage in an electrical circuit, the period changes to 2T, then 4T and so on. This type of bifurcation is known as 'period doubling'. In 1976, shortly after the Li and Yorke paper, Robert May of Princeton University wrote a paper in *Nature* which discussed how period doubling dynamics occurs in models in biological, economic and social sciences. He called the limit of these period doubling oscillations, *chaotic behavior*. Two years later in 1978, Mitchell Feigenbaum of Los Alamos National Laboratory wrote a paper on the mathematical properties of periodic doubling and its near universal existence in simple models of complex behavior of biological and physical systems. (see James Gleick's *Chaos* 1987, for a discussion of Feigenbaum's discovery.) Feigenbaum won the prodigious Wolf Prize in physics for his work (Fig. 6.3).

Shortly after these mathematical papers, experimental studies appeared in which all the dynamic phenomena of period doubling, chaos and strange attractors were observed in fluid, electrical, optical and mechanical systems, and the new science of Chaos Theory was born. In the decade that followed, attempts were made to explain all manner of unpredictable phenomena in physics and biology using the modern mathematics of nonlinear dynamics.

Like many technical and scientific fields, nonlinear dynamics and chaos theory bifurcated into different paths of research. There emerged several schools of ideas. Some of these included:

(i) The fractal school; complex geometry;
(ii) Bifurcation theorists; i.e. what are the paths to chaos?
(iii) Chaotic phenomena; strange attractors
(iv) Turbulence theory; the complex motions of fluids and weather
(v) Hamiltonian dynamics; a world of frictionless particles
(vi) Thermodynamics school; whence the origin of entropy or disorder?
(vii) Complexity theorists; self organization in biology and social science.

These schools of ideas have been discussed in the dozens of popular books on Chaos Theory and it would be a diversion of our task if we took them all up.

Since the late seventies, the ideas of chaotic and nonlinear dynamics have penetrated physics, chemistry, biology, economics, engineering and even financial policy and family therapy. Electrical engineers have developed chaotic circuits to produce secure data communication encryption technology and in Japan there is an official national committee on Chaos Engineering.

6.2 Tracing the Social Network for Chaos Theory

In some respect it is easier to track the links between modern scientists than links between earlier counterparts, say in aviation, because modern researchers cite references at the end of their research papers. However the first order problem is to identify the core set of nodes in early nonlinear dynamics and chaos theory. The problem for historians is that nonlinear dynamics was spawned in different fields of science and technology and each review of the subject lists its own core node set. For example, a review in 1983 by mathematicians John Guckenheimer and Philip Holmes, then of UC Santa Cruz and Cornell, respectively, listed few experimental physicists. (This book recently received an award in 2013, from the American Mathematical Society as a seminal work in the field for the last 30 years). Another review by the German H.G. Schuster in 1988, listed many experimental papers as well as mathematicians but did not list any papers originating from technology and engineering such as radio electronics and structural mechanics.

Using either Schuster's list of nodes or that of the mathematicians Guckenheimer and Holmes (1983) one can see an explosion in the number of nodes around 1980. For example in Fig. 6.4, middle graph, in 1975 there were around 25 players that had doubled in 1980 to over 50 nodes and in 1985 this figure tripled to more than 150 nodes. The data in Fig. 6.4 does not show the number of papers published, since an attempt was made to only count each research paper author once in the node count. The number of papers published in the early 1980's was far higher since many authors published as many as three or more papers each year. Also as we explained, Schuster did not survey the applied technology journals where many additional nodes and papers would be added to the total.

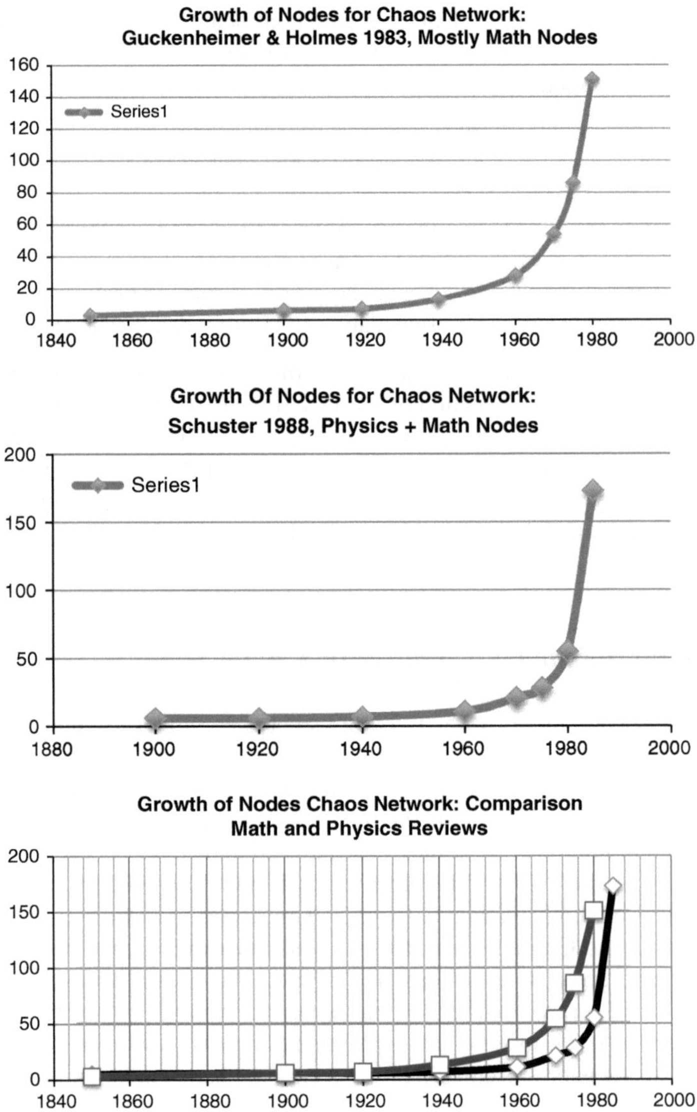

Fig. 6.4 Growth of number of research authors in the field of nonlinear dynamical systems and chaos theory as reviewed in (*top*) Guckenhiemer and Holmes (1983), (*middle*) Schuster (1988) and a comparison of these curves in the lower graph. -0The curve with data *squares* is based on the mathematics book

Thus it becomes clear, as in the case of the internal combustion engine, that once the sharp rise in the node growth curve is reached, it becomes virtually impossible to trace links between nodes in a historically accurate way. Instead we shall attempt to construct a partial or incomplete social network, principally between nodes that emerged by the 1970's, and then for a limited subset as we shall describe. Complete-

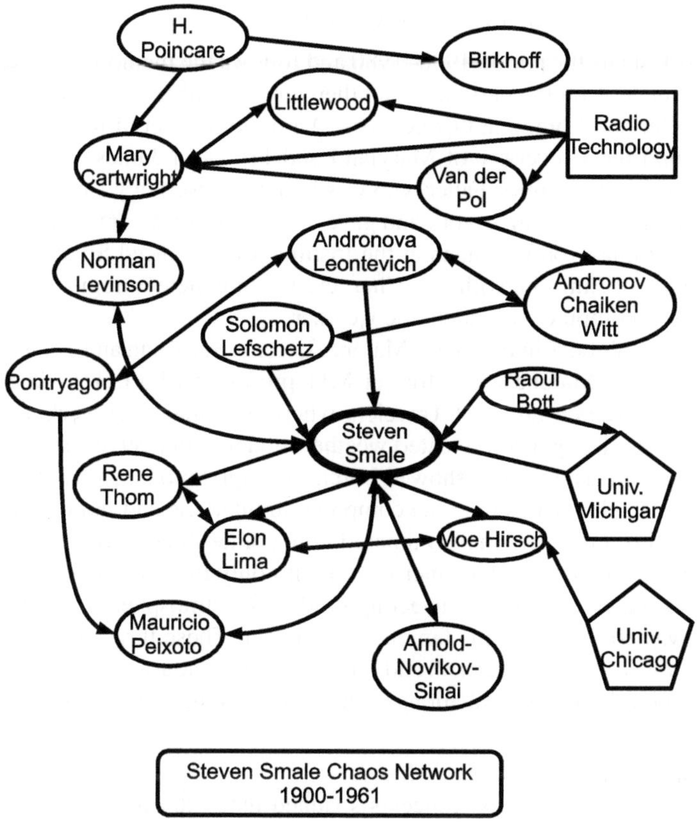

Fig. 6.5 Steven Smale's chaos network for nonlinear dynamical systems (Abraham 2000)

ness is not our goal here, but we wish to show that even in theoretical physics and mathematics, it is the network from which new ideas grow, promulgate and diffuse. Although mathematicians like to see their craft as different from engineers for example, the nature of human creativity is the same. Be it grimy engine innovation or lofty theoretical constructs, similar patterns appear.

The Standard Model Nonlinear Dynamics Network

Using a collection of memoir chapters edited by Abraham (2000), we have built a historical innovation network shown in Fig. 6.5 centered around the UC Berkeley mathematician Steven Smale for the period 1900–1960. This is necessarily a 'star' network of mostly mathematicians. The engineer van der Pol is included if only to connect to British mathematicians Mary Cartwright and John Littlewood who were intrigued with van der Pol's mathematical equations of electronic circuits. A collection of original papers on chaos by mathematicians and physicists before 1984 can be found in Hao (1984).

The Edward Lorenz-Fluid Mechanics Network

This network spans the period 1900–1980 and follows the memoir of Lorenz (1993) (Fig. 6.6). Smale is on the periphery and there are a number of meteorologists such as Hide, Faller and Lorenz in this network. An important set of nodes is the 'small world' around the University of Maryland and Professor James Yorke and Allen Faller. According to Lorenz, Faller who was in the Meteorology Department, called Yorke's attention to Lorenz's work and gave Yorke's group a copy of Lorenz's 1963 paper. Lorenz links von Neumann at the Institute of Advanced Study at Princeton, once the home of Einstein, with Jule Charney, a young meteorologist, in the attempt to use the new computers to model the weather.

Another set of nodes in Fig. 6.6 is Malone, Lorenz, Saltzman and Traveler's Insurance Co. Thomas Malone was a former MIT professor who Lorenz knew had set up a weather research group at Travelers where Saltzman was working. In 1961, researchers were using a complicated weather model with twelve variables. Lorenz met Saltzman at Travelers who showed Lorenz a simpler model with only seven variables. Saltzman showed Lorenz his computer simulations and noticed that four of the variables became very small compared to the other three. Edward Lorenz asked Saltzman if he could use the simpler model with only three variables incorporating temperature and velocity of a convecting fluid in the atmosphere From this simpler model he was able to write his famous 1963 paper. These three equations became known as "*The Lorenz Equations*" and hundreds if not thousands of research papers were published about these equations of fluid chaos in the decades of the 1980's and 1990's.

Lorenz-inspired Economic Chaos Models

Nonlinear dynamical systems concepts spanned many disciplines of the physical, biological, and social sciences. An example is the study of chaos in economics. One of the pioneers of this field was Richard Goodwin of Cambridge University who in the 1970's had proposed dynamic models to predict changes in prices and quantities of goods. He proposed a system of four first-order differential equations using the concept of an economic potential function in which greater quantities of goods will drive down prices and higher prices will depress demand for goods. In 1990, Goodwin published *Chaotic Economic Dynamics*, in which he refers to Lorenz's chaos models with the rhetorical statement; "*Why is economics like the weather?: because both are highly irregular if not chaotic, thus making prediction unreliable or even impossible*".

Goodwin also cited the work of the mathematician, George Birkhoff, who had inspired Edward Lorenz. Many other books and papers were written describing unpredictable economic behavior. Among them is a book, *Chaotic Dynamics: Theory and Applications to Economics* (1992) by Alfredo Medio a former Italian student of Goodwin. Medio uses many concepts from the nodes in the Smale Network (Fig. 6.5) and has borrowed figures (with citation) from books in physics, mechanical engineering and civil engineering including, Schuster (1988), Moon (1987), Thompson and Stewart (1986). Rosser (1999) has traced the work of Goodwin and other economists before the exponential rise of chaos theory, demonstrating an early evolution of nonlinear dynamics in economics. This cross-linking of disciplines was one of the

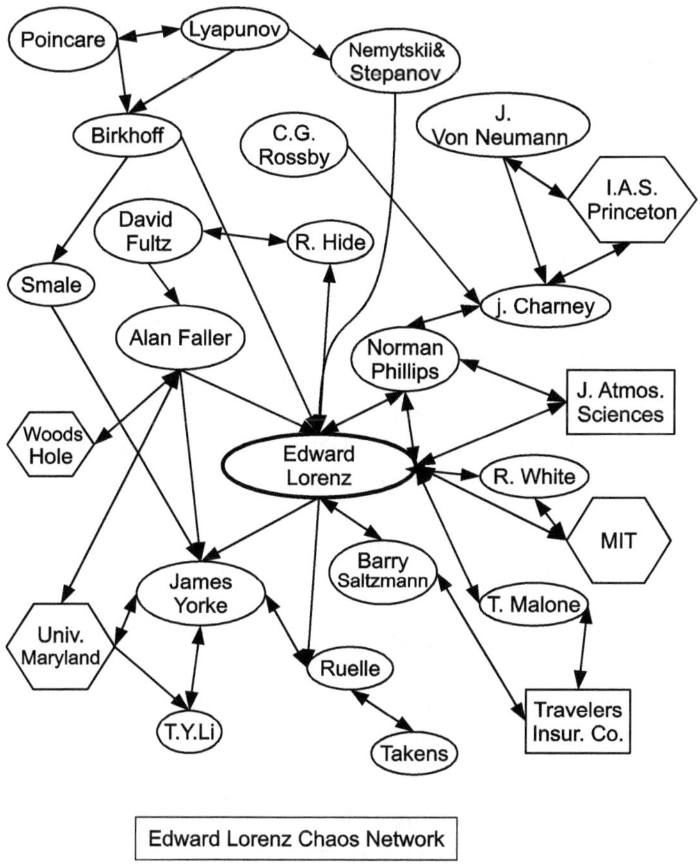

Fig. 6.6 Edward Lorenz weather, and fluid dynamics chaos network (Lorenz 1993)

drivers that produced the exponential rise in the number of nodes in the chaos theory network.

6.3 Radio and Engine Dynamics Origins of Chaos Theory

The mathematical basis of Chaos Theory is the study of nonlinear oscillations. Beginning with the periodic motion of the pendulum studied by Huygens in the seventeenth century, scientists and engineers began to catalog the variety of vibrations or oscillations in nature and technical systems. Elastic structures excited by internal combustion engines as well as steam turbines were observed to exhibit vibrations. Physicists and electrical engineers discovered that circuits could oscillate with such intensity that oscillatory electromagnetic signals could be generated which could be detected hundreds and thousands of miles away. In this way wireless telegraphy and radio emerged from the science of electrical oscillations.

Principal engineer-mathematicians and applied mathematicians who contributed to nonlinear dynamics in the pre-chaos era included:

Balthazar van der Pol [1889–1959] Dutch engineer, worked for Phillips on radio and electronics;
Georg Duffing [1861–1944] German engineer worked for Deutz engine manufacturer and Shell;
Stephen Timoshenko [1878–1972] Ukrainian-Russian engineer emigrated to US, worked for Westinghouse;
Jacob P. Den Hartog [1901–1989] Dutch engineer, emigrated to the US, worked for Westinghouse under Timoshenko, went to MIT;
Alexandr A. Andronov [1901–1952] Soviet era applied mathematician
Leonid Mandelsham [1879–1944] Soviet era mathematical physicist, worked for Karl F Braun in Germany
Richard Courant [1888–1972] German emigrant to US, worked with Hilbert, headed an applied mathematics institute in 1934 at NYU that studied fluids problems
Nikolai Minorsky [1885–1970] Russian emigre to US, worked on electronics problems
Chihiro Hayashi [1911–1987] Japanese professor of electrical engineering, Kyoto University, headed a large laboratory on nonlinear circuit dynamics

A large historical network diagram is shown in Fig. 6.7 spanning a century from 1880–1980. Beginning with Lord Rayleigh, mathematicians, scientists and engineers began a century long study of so-called nonlinear oscillations. Some of the major nodes in this network were Balthasar van der Pol of the Netherlands who worked for the Phillips company, famous for its electronics technology circa 1920. Van der Pol also worked for J A Fleming in England, who had invented a vacuum tube electronic device that would help revolutionize radio technology (Chap.5) Van der Pol also published a paper with Edward Appleton on nonlinear oscillations of triode vacuum tube circuits. Appleton went on to win the 1947 Nobel Prize in physics for his work on the ionosphere. At the time they were engaged in studying the propagation of radio waves in the atmosphere. This problem had earlier roots in Heaviside's attempts to explain Marconi's remarkable achievements in wireless telegraphy over extremely long distances (Stumpers 1960).

A social network map of the evolution of Duffing's and van der Pol's equations is shown in Fig. 6.7. One can see that there were several groups or clusters of nodes around the Westinghouse lab under Timoshenko, the Soviet Mandelst'ham-Andronov Institute, and the NYU-Princeton groups headed by Richard Courant and Solomon Lefschetz respectively.

The Radio Electronics-Engine Mechanics Sub-Network of nonlinear mathematics 1910–1980 shown in Fig. 6.7 emerges from at least two earlier technical networks—the engine dynamics of Otto and Diesel 1870–1900 and the wireless-radio electronics

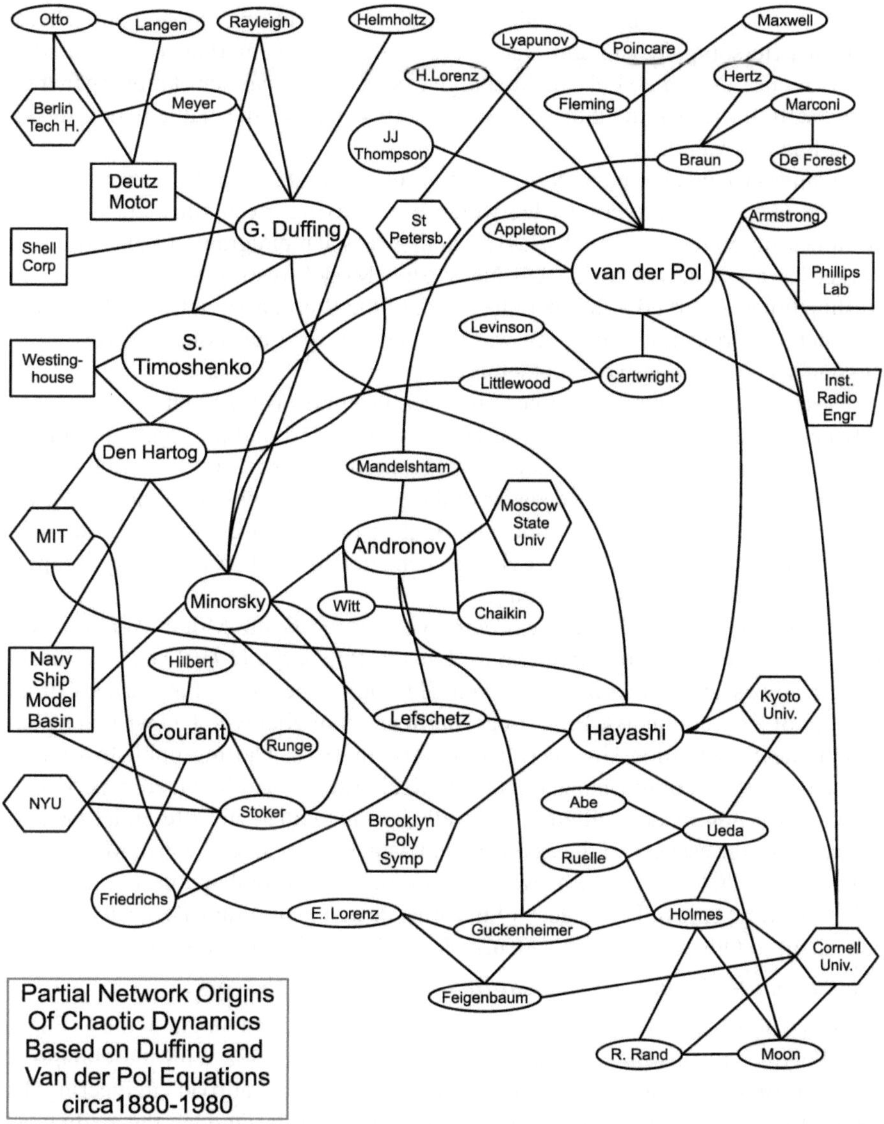

Fig. 6.7 Historical network for chaotic dynamics in engineering mathematics. Nodes and links are related to Dutch engineer Balthazar van der Pol and the electronics technology and to German engineer Georg Duffing and internal combustion engine technology

network of Marconi and de Forest, 1895–1925. Out of the work of Georg Duffing and Balthazar van der Pol, there evolved a number of sub-networks in applied nonlinear mathematics. This included the Russian-Soviet node cluster originating with Leonid Mandelstam in Moscow. Mandelshtam had studied with the German K F Braun who was the co-winner with Marconi of the Nobel Prize for wireless technology. During the 1920's and 30's, a number of talented immigrants including Stephen Timoshenko from St Petersburg Russia and Den Hartog from The Netherlands were hired by Westinghouse Corporate Labs in Pittsburgh to study the dynamics of motor generator systems. Both wrote influential books in the 1930's that used the mathematical models of Duffing and van der Pol.

Another group centered at NYU in New York around Richard Courant, included James Stoker and Kurt Friedrichs, who were studying the mathematics of water dynamics supported by the US Navy. Another Russian emigrant Nikolai Minorsky worked as a consultant to the US Navy Ship Model Testing Laboratory outside Washington also served as a link to the work of the Russian Soviet school. Soloman Lefschetz of Princeton University was another link for translating the work of the Russian schools of nonlinear mathematics.

These separate groups were joined in several symposia organized at the Brooklyn Polytechnic Institute in New York in 1953 and 1956, around the topic of nonlinear electric circuits. Not only did Lefschetz, Stoker, Minorsky and Friedrichs attend, but Chihoro Hayashi the future leader of the principal Japanese nonlinear EE group in Kyoto attended the 1956 symposium. Out of the Kyoto University group came Yoshisuke Ueda, who discovered chaotic behavior in 1961 in an analog computer study of a nonlinear electrical circuit.

In Germany, Georg Duffing was employed for the Deutz Motor Works in Koln and spent a short visit to Westinghouse Labs in Pittsburgh [1910] before returning to Germany as an engineering consultant. He studied the work of an earlier German inventor of a gyrocompass, Prof Oscar Martienssen of Kiel who had written a paper on nonlinear oscillations in electrical circuits. In 1918, Duffing published a book on the oscillations in mechanical and electrical systems and included experimental evidence for nonlinear oscillations as well as a mathematical differential equation that captures the essence of the experimental data. He was aided by Prof Meyer of the Berlin Technical University who was working on internal combustion engines. Berlin TH at that time enjoyed the legacy of machine design giant Franz Reuleaux, an early consultant to Otto and Langen's Deutz Motor Works (Kovacic and Brennan 2011).

The Russian Nodes in the Nonlinear Applied Dynamics Network

Stephen Timoshenko [1878–1972] was born in Ukraine, and taught at the St Petersburg Polytechnic Institute. He emigrated to the US in 1922 with the collapse of Imperial Russia. In Russia he had had some experience in railroad engineering. But his real love was the application of mathematics to structural engineering, particularly problems of elastic stability and vibrations of structures and machines. In Europe he traveled to Germany and Cambridge and met Ludwig Prandtl, and Theodore von Karman, both giants in the study of fluid mechanics.

In the US, he began work at the Westinghouse Laboratories in Pittsburgh. A few years later he hired a young student J Den Hartog from the Netherlands as his assistant in the study of dynamic problems in rotating machinery associated with the production of electrical generation technology. Timoshenko left Westinghouse to become a professor at University of Michigan and published his seminal book on mechanical vibrations in 1928. In it he cited the work of Duffing and his famous equation of nonlinear dynamics. A few years later, Den Hartog took a position at Harvard University (and later MIT) and published a similar book on mechanical vibrations and cited the work of van der Pol on nonlinear oscillations.

Timoshenko's 1928 book not only dealt with linear systems, but also "non-harmonic vibrations", using the dynamics of locomotives as an example. He also dealt with vibrations of turbine blades and described several vibration measuring instruments including seismic vibrographs. The 1948 book *Advanced Dynamics* of S. Timoshenko and D.H. Young, both at Stanford University at the time, combined advanced analytical methodology with many problems of direct machine dynamics. This includes balancing of reciprocating engines, dynamics of rotating ball governors, dynamics of constrained systems including linkages, flywheel governors, gyroscopic motions, the gyrocompass, ship stabilizers, and vibration absorbers. Analytical techniques included linear systems and nonlinear vibrations. The Authors acknowledged their access to several European books (Moon 2008).

Through Timoshenko's life, we can see a network of connections in areas of engineering dynamics and structural stability stretching from Russia to Germany, England and the US. These links were important not only in the nonlinear dynamics and chaos network of Fig. 6.7, but were also influential in the network discussed below related to catastrophe theory and structural stability.

Another Russian Mandelsh'tam, in Moscow State University, began to build a group to study nonlinear dynamics that included Andronov, Chaikin and Witt who in the 1930's published a seminal work on mathematical methods in the analysis of nonlinear oscillations. After World War II, Solomon Lefschetz of Princeton University would help translate these Soviet works into English. By this time two of the principal equations that captured many electrical and mechanical nonlinear phenomena were recognized as *the Duffing equation* and the *van der Pol equations.* Once in the mathematical literature, they began to be studied for their own sake and often the origins of these mathematical models were ignored or forgotten.

Lefschetz was also part of the conferences on nonlinear oscillations in electrical circuits held at the Polytechnic Institute of Brooklyn, in 1953, 1956, that included another Russian ex patriot Nicolai Minorsky. He and another attendee, James Stoker of NYU, wrote important textbooks that would help enshrine the nonlinear models of Duffing and van der Pol. Minorsky became a consultant to the US Navy in the study of stabilization of ship oscillations at sea.

It is interesting to note that this network is linked to Albert Einstein through Professor Oscar Martienssen of Kiel while Einstein was at the patent office and involved in a dispute about the gyrocompass of Elmer Sperry in the US and Anshutz in Germany. It is also noteworthy that in the standard model of the origins of nonlinear dynamics and chaos theory, (Fig. 6.4), key links such as Timoshenko, and Den Hartog are often ignored.

Another important cluster to emerge after World War II, was the Japanese nonlinear dynamics group at Kyoto University under the leadership of Professor Chihiro Hayashi. He had studied the works of Lord Rayleigh, Poincare', Liapunov, Duffing (1918), and Van der Pol. (Ueda 2001) Hayashi can be traced as a visitor to MIT and the Brooklyn Poly nonlinear circuit symposium in 1956. He later visited Columbia University and Cornell University around 1966.

Personal Connections to the Network

The author was fortunate to attend Professor Hayashi's lecture at Cornell University in the spring of 1966 as a graduate student in Theoretical and Applied Mechanics. Hayashi's student, Yoshisuke Ueda, now retired from Kyoto University, would become famous for discovering the "*Japanese Attractor*" in chaotic dynamics (1961), based on Duffing's equation, using analog computer studies of electrical circuits. (see e.g. Ueda 2001). The author also met Professor Ueda in 1978 in Kyoto Japan and later hosted him in Ithaca NY. In the 1970's the author was present at a lecture of Den Hartog when he received the Timoshenko Medal of the American Society of Mechanical Engineers. At Princeton the author attended the last lectures of Solomon Lefschetz in the 1960's as well as a lecture by Richard Courant of NYU. Later he was inspired by the work of Philip Holmes of Cornell (now at Princeton) in the late 1970's and Michael Feigenbaum who lectured at Cornell in the early 1980's. The textbooks of Steven Timoshenko were popular among graduate students in the 1960's as were the books of Minorsky and Stoker. Thus when the ideas of chaos theory were nascent in the 1960's, the author was primed by his prior connections to some of the early leaders of applied nonlinear dynamics when he began his own research in this subject in 1976 (see e.g. Moon 1987, 1992).

Many nonlinear science groups that emerged in the 1970's and 1980's are not shown in the network diagram of Fig. 6.7, including the Santa Cruz California group (e.g. John Guckenheimer), the Virginia Tech group under Ali Nayfeh, the University College group in London under J.M.T. Thompson and the University of Maryland group around James Yorke, Edward Ott and Celso Grebogi. A UC Berkeley node would include Jerald Marsden (math), Leon Chua (EE) and C-S. Hsu (ME). The choice of nodes in Fig. 6.7 was made to show how the origins of nonlinear science and mathematics had roots in the technical engineering networks of the late nineteenth and early twentieth centuries as described in Chaps. 2, 3, 6.

Up until the 1970's, the dynamic theories of machines viewed the machine as a deterministic entity whose behavior could be uniquely predicted and controlled by use of the Newton-Euler laws of motion. (See e.g. Moon 2008) Most machines with rotating parts were assumed to behave in a periodic manner. Unpredictable noise was usually assumed to come from outside random agents such as fluid turbulence, or road or track irregularities in the case of land vehicles. In his 1875 book on Kinematics of machines, Reuleaux (Chap.2) recognized the problem of unpredictable noise in machines. Twentieth century authors in technical dynamics, such as Den Hartog, studied so called self-excited motions in machines that usually generated nonlinear periodic limit cycle motion of the type studied by van der Pol.

Modern research has shown that many machine mechanisms can exhibit small amounts of unpredictable or chaotic dynamics due to the inevitable imperfect nature of the machine as constructed, including friction, backlash, and elastic flexibility. Examples include chaotic noise in gears and chaos in ball bearings as well as linkages with elastic elements or loose joints (Fig. 6.2). This has suggested that a modern theory of machines should admit a certain measure of self-generated, chaotic noise in the behavior of mechanisms.

Another area of unpredictability is the stability of engineering structures. Thin structures under loads are known to collapse even when the material stresses are not high. Shell-like structures are especially vulnerable under compressive loads as the reader can experiment with an empty drink can. The study of buckling of structures begun by Timoshenko and others a century ago led in the 1970's to another mathematical study called *catastrophe theory* that laid some of the foundation for chaos theory in the 1980's as we discuss below (Fig. 6.8).

6.4 Catastrophe Theory and Chaos 1960–1980

Before the media hype for chaos theory in the late 1980's the mathematical magic of the 1970's was *catastrophe theory*. Similar to fractals and chaos theory, it was a pictorial, graphical approach to modeling a wide range of problems from biology to collapsing bridges. In April 1976 it captured the front cover of Scientific American with an article by E. Christopher Zeeman of the University of Warwick UK, (Zeeman 1977). Its French connection and champion was the mathematician Rene' Thom who in the preface to his book traced his interest to D'Arcy Thompson's 1917 classic work in biology, *On Growth and Form.* Parallel to this mathematical development was an engineering research area into the collapse of metal structures. (See e.g. Fig. 6.8.) This had an active group in the UK and the two areas of catastrophe theory and buckling of structures seemed to find a resonance in several UK universities. A partial network for catastrophe theory and stability of structures in the UK and elsewhere is shown in Fig. 6.9. The nodes in Fig. 6.9, shown in bold, have UK affiliations.

Instabilities in Engineering Systems

In several areas of classical physics it was known that material systems could become unstable under certain load conditions. In the case of aviation, one of the most important problems was the matter of stability of the aircraft, a problem that the Wright brothers understood and solved brilliantly. In the case of electron flow between electrodes as in early wireless telegraph systems it was the critical voltage for which electric current will begin to flow across the electrodes.

Structural engineering was one of the earliest problems where mathematics of stability was applied in the nineteenth century. In Fig. 6.8 is shown a schematic of a vertical elastic strut under compressive load. (Take a playing card between the thumb and middle finger and squeeze and the card will assume a curved shape. This is termed *'buckling'*.) The problem of buckling of thin rod-like structures was solved in the nineteenth century. However elastic stability of shell structures such as aircraft fuselages and rocket casings remained a challenging problem into the middle of the twentieth century.

Centers for the study of structural stability appeared in Germany, Holland, Russia and the UK. Structural analysts discovered mathematical properties of these systems

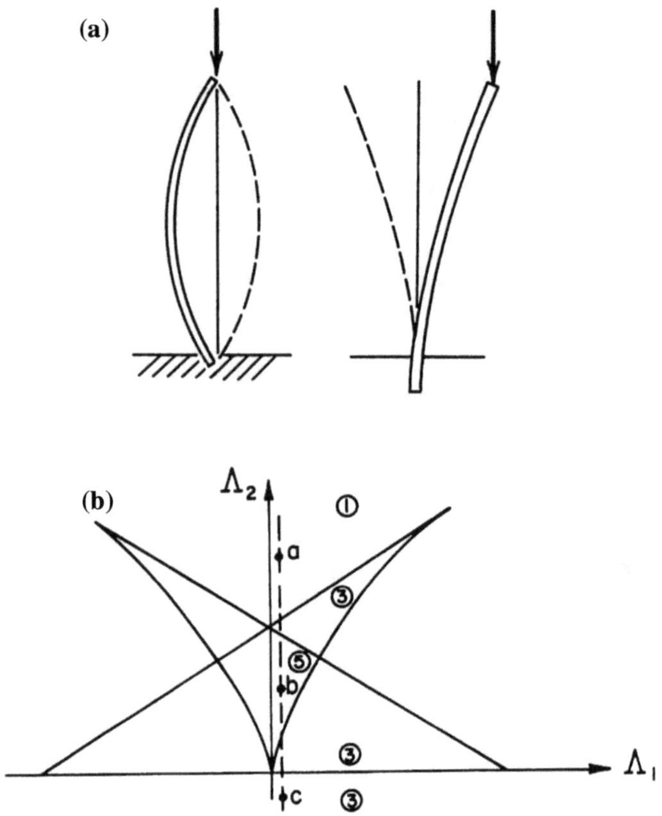

Fig. 6.8 Structural buckling and catastrophe theory. **a** Sketch of two elastic columns bucking under load. **b** Butterfly catastrophe. The vertical scale represents the load in (**a**) and the horizontal scale represents some geometric property of the structure. The figure shows the regions where as the load goes from **a**–**c**, the number of possible equilibrium positions goes from 1 to 3 to 5 to 3. The boundary shape with its three cusps demonstrates the extreme sensitivity as to what state the structure will go under changes in the loads and geometry

that mathematicians later found to be identical to those in the field of catastrophe theory. Some of the connections between these groups are shown in the network shown in Fig. 6.9. Mathematicians such as Rene Thom, Arnold and Zeeman, were brought together with structural engineers such as Chilton, Thompson, Hunt and others. These nodes are further linked to Timoshenko in Russia. Thompson traces part of his links to structural stability to Koiter [c. 1945] and Biezeno [c. 1928] in Holland. Koiter and Biezeno were motivated to study buckling problems by attempts to minimize weight in aircraft design.

Catastrophe theory provided a mathematical methodology to classify the number of equilibrium positions of the structure as a function of loads and geometric parameters such as small deviations from perfect geometry. An example is shown in Fig. 6.8 bottom. This figure shows a multi-cusped shape denoted as a *'butterfly' catastrophe*

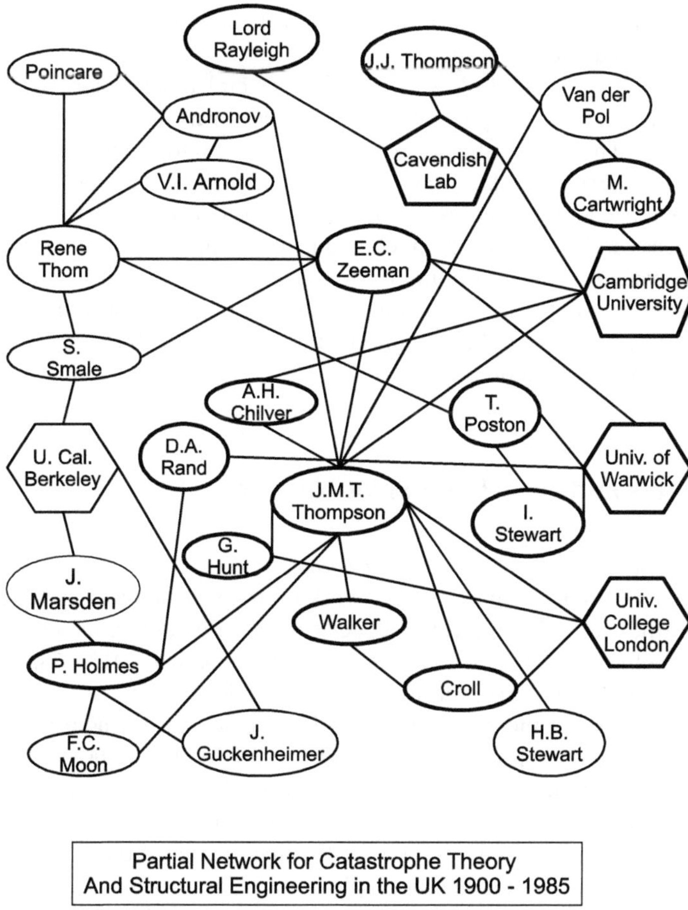

Fig. 6.9 Partial social network for catastrophe theory and engineering structural stability in the United Kingdom circa 1980. The *bold nodes* are associated with the UK. This network overlaps with the Smale chaos network Fig. 6.5 and the van der Pol Network in Fig. 6.7

for its wing-like appearance. Every point in the plane denotes a load parameter and a geometric parameter. The numbers in the separate regions of the 'butterfly' indicate the possible number of equilibrium positions the structure could take. Multiple equilibria meant that the solution was not unique and there was unpredictability as to how the structure would behave. This coupling of stability and unpredictability was analogous to the unpredictability to be found in dynamic problems of chaos.

Many of the players in the catastrophe theory and elastic stability network in Fig. 6.9 did not move into chaotic dynamics. Michael Thompson of University College London built a large and influential group that interacted with other nonlinear dynamics laboratories such as those at Cornell, that included the author's experimental dynamics group as well as well-known theoreticians Philip Holmes, John Guckenheimer and Richard Rand.

During the 1970's and 1980's, there were several other schools of nonlinear dynamics related to chaotic phenomena. One school was interested in the '*road to chaos'*—how did normal predictable dynamics change into chaotic or unpredictable dynamics? This school was known as *bifurcation theory* (See e.g. Nayfeh and Mook 1979). Another school, of interest to physicists, was concerned with the full blown chaotic process and sought to find mathematical measures to characterize the chaos in physical systems. This school might be called the *fractal chaos* school. And the designated genius-innovator was Benoit Mandelbrot of IBM. But similar to catastrophe theory, fractal theory did not begin with dynamics and chaos but with geology and the mathematics of geometric shapes.

6.5 Fractals and the Mandelbrot Network

Benoit B. Mandelbrot: the "Father of Fractals"?

Just when we think an innovation network is complete, another historical source appears. In the case of nonlinear science it was the autobiographical memoir of Benoit Mandelbrot [1924–2010], published posthumously in 2012. Mandelbrot was awarded the 2003 Japan Prize for his work on fractals.

Fractals was one of the defining hallmarks of chaos theory as it became a tool to characterize the complexity and the hidden order in chaotic processes (See e.g. Figs. 6.10 and 6.11). In the standard history of fractal mathematics there was a long list of earlier nodes in the fractal network (Fig. 6.12) Cantor, von Koch, Sierpinski, Julia, Hausdorff were all important before the first quarter of the twentieth century. When reading Mandelbrot's autobiography, one is immediately struck by the fact that he came from a family and educational background that had direct links to these earlier 'fractalists'. (Seirpinski a Polish mathematician, had advised Benoit's uncle Szolem Mandelbrot to take a position in a French University that later saved the life of the Polish Jewish Mandelbrot family in the wake of the Holocaust.)

Despite the desire of Benoit's uncle Szolem to see him in the arena of abstract mathematics, the younger Mandlebrot 'rebelled' first by enrolling at the more engineering focused Ecole Polytechnique and then by pursuing applied mathematics. A perusal of his published papers finds interests in statistical economics and thermodynamics, communication and information theory, geography and hydrology and fluid turbulence. He was not afraid to take empirical data, such as the word frequency data of Zipf (1949) or turbulence pressure measurements and find fractal measures in them. However we do not see his explicit interest in nonlinear dynamics as found in the van der Pol equation of vacuum tubes or the Duffing equations of engine dynamics. It remained for others to take his ideas about fractal sets from the mathematics of complex numbers or the geographers measure of the coastline of Britain and apply them to all manner of chaotic dynamics in physics and engineering.

The connections between chaos and fractals did not fully emerge until about 1980–1985 by which time the exponential growth of nodes in the innovation network

Fig. 6.10 Fractal set of 100,000 points related to the chaotic dynamics of an electrical circuit generated in the Laboratory of Yoshisuke Ueda, Kyoto University, c. 1980. A zoom into any part of this fractal reveals further structure

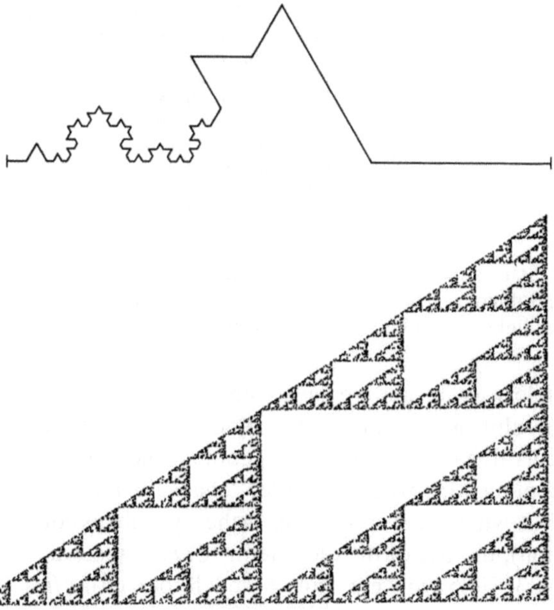

Fig. 6.11 Fractal sets: *Top* Koch curve constructed by adding ever smaller triangular elements. *Bottom* Sierpinski triangle; constructed by subtracting smaller and smaller triangular elements. Sierpinski was a family friend of Mandlebrot's relatives in Poland

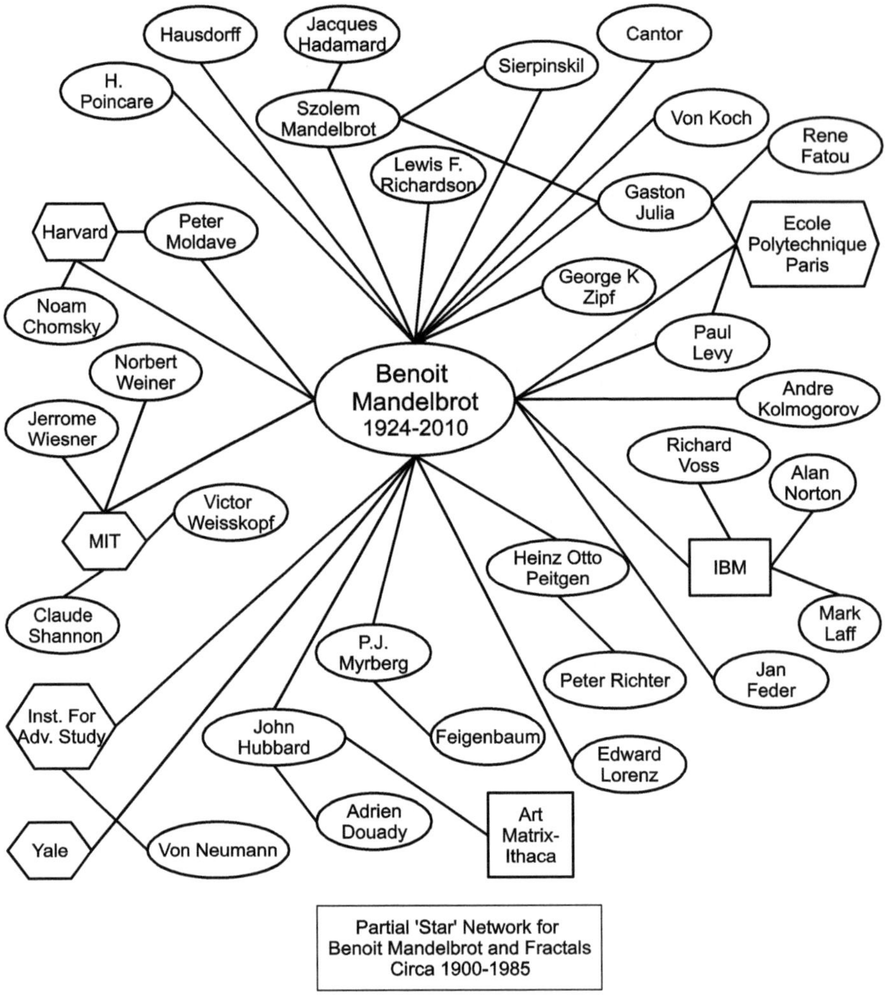

Fig. 6.12 Partial 'Star' Network for Benoit Mandelbrot and the field of fractals, based primarily on Mandelbrot's posthumous autobiography *The Fractalist*, published in 2012. This star network neglects most cross links, except those with the Mandelbrot node

were already underway. Just as we have seen the merging of electronics and engine networks with nonlinear dynamics in the 1950's–1960's, we see the merging of the nonlinear systems dynamics and fractal geometry networks beginning in the late 1970's and becoming what would be called 'chaos theory'.

In the author's own work in applied chaos theory with his colleague Philip Holmes (now at Princeton), the use of the older term *'Cantor set'* was used in papers circa 1980, but by 1983–1985 we were using Mandelbrot's term *'fractal'* to describe the strange pictures of chaotic dynamics that we were observing in our laboratory and computer simulations. (See Fig. 6.13). This was no doubt in part because of the writings of Mandelbrot but also because of the exponential tide of research papers in fractal geometry that his earlier books of 1977 and 1982 had triggered.

Without taking anything from Mandelbrot's genius, in reading his memoir with the history of mathematics as a background, one feels that he was destined to define the field of fractal geometry, just when the field of nonlinear dynamics and chaos needed it. In the same way in which the Lorenz network at first seems almost unconnected to the "standard Model" of nonlinear dynamical systems of Smale, Mandelbrot's fractal network seems disjoint from that of van der Pol and Duffing until they merge in the 1970's and 1980's. A similar convergence of networks could be drawn in the area of computer graphics and the development of computer hardware at IBM. It is no coincidence that Mandelbrot worked for IBM and had access to computer programming expertise. In picture credits and acknowledgements he lists several IBMers such as Richard Voss and Mark Laff.

In mathematical computer graphics the name of Cornell mathematician John Hubbard often appears as someone who delved into the mysteries of the so-called "Mandelbrot set", and by some accounts, actually named the mathematical objects that resulted from the computer iterations after Mandelbrot. But Hubbard's name does not appear in either the 'Smale standard model' or Mandelbrot's history of fractals. No matter how thorough one is about constructing innovation networks, it is likely that there will always be nodes that are missed or hidden from the official histories, such as critical technicians, programmers, or colleagues who made an important suggestion. The innovation network is only a crude approximation to how human societies create new ideas and technologies.

Fig. 6.13 Fractal Chaos. Fractal patterns in the study of the dynamics of oscillating pendulums and buckled elastic structures produced in the author's laboratory at Cornell University in the 1980's (Moon 1992)

This chapter closes with the observation that intersecting networks that contribute to a given scientific or technical field are so complex that it becomes almost impossible to represent it in a book format of this kind. One is reminded of artist Mark Lombardi who used a wall format (3 m by 1 m] to represent the interlocking links between dozens of American companies and individual directors. See e.g. the work of the late artist Mark Lombardi [1951–2000] on the website of the Brooklyn art gallery *Pierogi,* in New York at http://www.pierogi2000.com/artists/mark-lombardi/.

In the case of nonlinear science and mathematics of the late twentieth century, such a complex network would encompass the telegraph and electronics networks, internal combustion engine and airplane dynamics networks, meteorology and fluid dynamics networks, applied analysis and fractal sets networks in mathematics. Such a dizzying array would certainly cover a wall similar to that of the late Lombardy.

6.6 Summary

From these five bewildering network diagrams and litanies of unrecognizable names what can one take away from this chapter? First in looking at Figs. 6.1 and 6.4, the diffusion of these ideas into the cultural landscape lagged behind the rise in the growth in the number of players in the network. Another idea that has been noted several times in this chapter is the interrelationship between mathematical science and technology. In the case of chaos theory this dependency was not only related to the substance of chaos theory such as the mathematical equations related to machines and electronics but also the tools that mathematicians used, namely computers that became essential to the mathematical methodology.

Edward Lorenz described the early efforts to model the weather with computers in the 1950's. His own work used computers to explore the complex dynamics of the mathematical models. Certainly the study of fractals by Benoit Mandelbrot would not have advanced as it did without his access to IBM's expertise. And the beautiful fractal pictures of John Hubbard of Cornell relied on new computer equipment in the engineering college.

Another result of examining the networks of chaos theory is the appearance of multiple sub-networks. Unlike technologies, such as the internal combustion engines, where all the players were focused on the same type of artifact, in the case of mathematical theories many of the players were unaware of the other sub-networks. Even in the same generation, many of those who studied the stability of engineering structures and developed an interest in catastrophe theory, never had any interest in chaotic dynamics a decade later. Many mathematicians working on Smale's 'horseshoe mapping theory' of chaos , had no interest in fractal theory. Whereas most of the experimental physicists testing the predictions of chaos theory in their laboratories took a serious interest in the theory, many theorists were oblivious to experimental work in chaos theory. Yet in spite of these disparate groups, there were players who crossed the boundaries and ventured into the other sub-networks such as Zeeman at Warwick in catastrophe theory, Jim Yorke at Maryland who delved into Lorenz's weather dynamics equations and Philip Holmes, then at Cornell, who brought together structural dynamics and Smale's mathematical theorems.

What is also remarkable for human culture in the late twentieth century, is that although there was no one genius or hero in chaos theory who brought all the pieces together, the network produced a body of knowledge that became enmeshed in the fabric of modern physical science and mathematics.

Epiphany Stories in Chaos Theory

Following James Watt and his steam engine condenser (Chap.1) or Willis Carrier and his air conditioner (Chap.7) some of the principals in Chaos Theory have recalled their own epiphany stories. Steven Smale wrote of his getting some of his ideas while he walked the beach in Rio de Janeiro. Edward Lorenz talked about visiting a colleague at an insurance company research center and discovering the clues to a simpler model for weather dynamics. Yoshisuki Ueda has written of his excitement as a graduate student discovering strange signals from an analog computer in 1961 and then struggling to convince a skeptical research advisor that the data were correct. No one says to themselves "I am part of a large historical social network that has inspired my discoveries." Everyone's experience of discovery and creativity is unique. Yet when looking at the big picture, be it radio physics or chaos theory, one cannot ignore the milieu in which we all work, our education, colleagues, technical equipment, books and papers that we read and probably most of all the information commons that surrounds and engulfs us.

References

Abraham R (2000) The chaos Avant-Garde. World Scientific Publishing Co, Singapore
Andronov AA, Vitt AA, Khaikin SE (1966) Theory of oscillators. Pergamon Press, Oxford
Arnold VI (1984) Catastrophe theory. Springer Verlag, Berlin
Britvec SJ (1973) The stability of elastic systems. Pergamon Press, New York
Croll JG (1972) Elements of structural stability. Wiley, New York
Diacu F, Holmes P (1996) Celestial encounters. Princeton University Press, Princeton
Duffing G (1918) Erzwungene Schwingungen bei veränderlicher Eigenfrequenz und irhe technische Bedeutung. Druck und Verlag von F. Vieweg & Son, Braunschweig
Feder J (1988) Fractals. Plenum Press, New York
Gleick J (1987) Chaos: the making of a new science. Viking Penguin, New York
Goodwin RM (1990) Chaotic economic dynamics. Clarendon Press, Oxford
Guckenheimer J, Holmes P (1983) Nonlinear oscillations, dynamical systems, and bifurcations of vector fields. Springer-Verlag, New York
Hao B-L (1984) Chaos. World Scientific Publishers, Singapore
Hayashi C (1964) Nonlinear oscillations in physical systems. Princeton University Press, Princeton
Holmes PJ (1990) Poincare celestial mechanics; dynamical systems theory and 'Chaos'. Phys Rep 193(3):138–163
Kovacic I, Brennan MJ (eds) (2011) The duffing equation: nonlinear oscillators and their behavior. Wiley, New York
Li T-Y, Yorke J (1975) Period three implies Chaos. Am Math Mon 82:985
Lorenz E (1993) The essence of Chaos. University of Washington Press, Seattle
Mandelbrot BB (1977) Fractals form chance and dimension. W.H. Freeman and Co, New York

Mandlebrot B (1982) The fractal geometry of nature. W.H. Freeman, San Francisco
Mandelbrot B (2012) The fractalist: memoir of a scientific maverick. Pantheon Books, New York
Medio A (1992) Chaotic dynamics: theory and applications to economics. Cambridge University Press, Cambridge
Minorsky N (1962) Nonlinear oscillations. D. Van Nostrand, Princeton
Moon FC (1987, 2004) Chaotic vibrations. Wiley, New York
Moon FC (1992) Chaotic and fractal dynamics. Wiley, New York
Moon FC (2008) History of dynamics of machines and mechanisms from Leonardo to Timoshenko. In: Yan H-S, Caccarelli M (eds) International symposium on history of machines and mechanisms. Springer, pp 1–20
Nayfeh AH, Mook DT (1979) Nonlinear oscillations. Wiley, New York
Peitgen HO, Richter PH (1986) The beauty of fractals. Springer Verlag, Berlin
Poston T, Stewart IN (1976) Taylor expansions and catastrophes. Pitman Publishing, London
Rosser JB (1999) The pre-history of chaotic economic dynamics. In: Sertel MR (eds) Contemporary economic issues. Macmillan, vol 4, pp 207–224
Ruelle D, Takens F (1971) On the nature of turbulence. Comm Math Phys 20:167, 23:343
Schroeder M (1991) Fractals, Chaos power laws. W.H. Freeman & Co, New York
Schuster (1988) Deterministic Chaos. VCH Verlag, Weinheim
Stoker JJ (1950) Nonlinear vibrations. Interscience Publishers, New York
Stumpers FLHM (1960) Bath. Van der Pol's work on nonlinear circuits. IRE Trans Circ Theor 7(4):366–367
Thom R (1975) Structural stability and morphogenesis. W.A. Benjamin Inc., Reading
Thompson DW (1917) On growth and form. Cambridge University Press, London
Thompson JMT (1973) A general theory of elastic stability. Wiley, London
Thompson JMT (1982) Instabilities and catastrophes in science and engineering. Wiley, New York
Thompson JMT, Stewart HB (1986) Nonlinear dynamics and chaos. Wiley, New York
Timoshenko S (1928) Vibration problems in engineering. D. van Nostrand, New York
Timoshenko S, Young DH (1948) Advanced dynamics. McGraw-Hill, New York
Turcotte D (1992) Fractals and Chaos in geology and geophysics. Cambridge University Press, Cambridge
Ueda Y (1992, 2001) The road to Chaos-II. Aerial Press, Santa Cruz
Van der Pol B (1960) Selected scientific papers. In: Bremmer H, Bouwkamp CJ (eds) vol 2. North-Holland Publishing Co., Amsterdam
Van der Pol B, Van der Mark J (1927) Frequency demultiplication. Nature 120(3019):363–364
Zeeman EC (1977) Catastrophe theory: selected papers, 1972–1977. Addison-Wesley Publishing Co, Reading
Zipf GK (1949) Human behavior and the principle of least effort: an introduction to human ecology. Addison-Wesley Press, Inc., Cambridge

Chapter 7
The Last Hero-Inventor: A Theory of Innovation?

Abstract In our last iconoclastic example of the genius-inventor we construct a social network for Willis Carrier and the beginnings of air conditioning technology. In a summary of exponential knowledge growth in other technologies, we draw the conclusion that human patterns of innovation have been similar across two centuries but that the time scale for explosive growth has been decreasing dramatically. We also review the elements for constructing social networks in the history of technology and speculate on models for innovation.

From clocks in the eighteenth century to chaos theory in the twentieth century we have charted the connections between history's knighted hero-inventors and geniuses and their sometimes hidden networks through which they found their epiphany moments. Characteristic of these examples has been evidence for exponential increases in events and product acceptance accompanied by an increase in the historical innovation nodes and links in the network. We have provided evidence that the statistical nature of these networks has characteristics of contemporary social networks and the World-Wide Web. Do a few examples make a universal theory? Of course not. As an experimentalist the author was led from one case study to the next in search of exceptions. Perhaps one exception is the ballooning technology of the late eighteenth to the early twentieth century, a case of linear progress ending in the failure of Count Zepplin's dirigible.

In the twentieth century, the technology of magnetic levitation of trains was another example of intense public interest and engineering excitement in the 1970s and 1980s, only to falter as conventional rail closed the speed gap in the 1990s. In physics there is the case of cold fusion in the 1990s. It is not the claim that only successful technical and scientific innovations arise from complex historical networks, but that both enduring and fading innovations arise from social networks. Whether these innovations become embedded in modern society depends on many factors independent of the inherent nature of the innovation or invention. Our point is that innovations and inventions by their nature demand a societal network to come before

F. C. Moon, *Social Networks in the History of Innovation and Invention*,
History of Mechanism and Machine Science 22, DOI: 10.1007/978-94-007-7528-2_7,

© Springer Science+Business Media Dordrecht 2014

the stage of history. Economic, political and societal factors cast the ultimate votes for acceptance.

Sometimes the failed technology, such as steam piston engines, enables a newer, more enduring one. It set the stage for the Industrial Revolution only to be replaced by the steam turbine used in power plants today and the internal combustion engine. Was the steam piston engine a failure? Most historians would argue not—it was a necessary step in the evolution of steam power and energy producing machines. Similarly with the decline of vacuum tube technology and analog electronics, only to be replaced by transistor electronics and eventually silicon-embedded chip circuitry and digital electronics. Could silicon-based electrical products have leap-frogged de Forest's triode tube and Armstrong's superheterodyne circuits? Evolution in biology took many branched paths and technical and scientific evolution is probably no different.

Before becoming too pontifical in this summary chapter, we look at one more example of a technology that was successful a century ago, survived and changed the demographics of living patterns on Earth, and threatens to be a key factor in solving the threats to global climate change. This technology is air-conditioning and the heir apparent to the hero-inventor in Willis Carrier, another Cornell University mechanical engineering graduate of 1902.

Fig. 7.1 Willis Carrier c. 1915, cornell class of 1902, mechanical engineering. Introduced humidity control to the heating and cooling of buildings

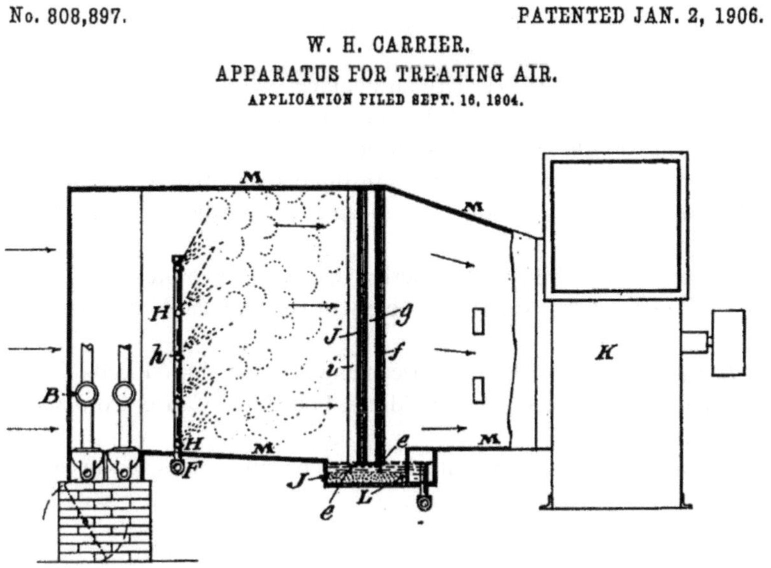

Fig. 7.2 Willis Carrier first air conditioning patent 1906, No. 808897

7.1 Willis H. Carrier and the Persistence of the Genius Myth

In a recent book on innovation by Johnson (2010), *Where Good Ideas Come From*, Johnson makes a very strong case for the importance of societal parameters for the creation and diffusion of innovative ideas, including social networks. However in his last chapter Johnson tries to demonstrate that even though most ideas are evolutionary and collaborative, there are a few that come from the individual genius. His principal example is Willis Carrier [1876–1950], Fig. 7.1, and the invention of air conditioning (See patent front page in Fig. 7.2). There was no intent to study this class of inventions because the ancillary subject of thermodynamics is broad and deep with many important players spread out over a century or more. But the challenge was there to see if Carrier had created modern air conditioning machines in the absence of a social network.

The principal nodes in air conditioning technology and related thermodynamic science are listed below.

Principal Nodes in Applied Thermodynamics Network

Sadi Carnot [1796–1832] French engineer who proposed limits to heat engines.
Rudolf Clausius [1822–1888] German physicist. Put Carnot's ideas into formal theory. Defined the concept of entropy.
William F. Rankine [1820–1872] Scottish physicist and engineer. Described the basic laws of thermodynamics.

Lord Kelvin [1824–1907] British engineer and physicist. Contributed to science of electricity and thermodynamics.
Carl von Linde [1842–1934] German engineer who pioneered in refrigeration technology. Teachers at Swiss ETH were Clausius and Reuleaux.
Robert H. Thurston [1839–1903] American mechanical engineer, expert on the steam engine, and professor at Stevens Institute and Cornell University.
Alfred R. Wolff nineteenth century. American engineer who introduced heating and cooling of buildings.
Rolla C. Carpenter [1852–1919] American engineer and professor at Cornell University. President of the American Society of Heating and Ventilating Engineers 1898.
Willis H. Carrier [1876–1950] American engineer who introduced humidity control the ventilation and heating of buildings. Founder of Carrier Corp. Graduate of Cornell University.
Wendt Brothers Buffalo NY manufacturers of centrifugal fans and air-conditioning equipment in early twentieth century.

7.1.1 Air-Conditioning Machines Social Network

In Fig. 7.3 is a star network centered around Willis Carrier and gleaned from his biographies and web-based sources (Ingels 1952; Wampler 1989). Many of the important nodes in thermodynamics are not shown in this network.
Description of Carrier Network Links

Willis Carrier grew up on a farm in northwestern New York State near Buffalo and won a state scholarship to attend the engineering college at Cornell University, known as the Sibley College of Mechanical Engineering and Mechanics Arts in Ithaca New York. After graduation in 1902, he was hired by the Buffalo Forge Company which produced pumps and centrifugal ventilation equipment. At the same time the Sackett-Wilhelm Lithography Company in Brooklyn NY, wrote to Buffalo Forge to see if they could purchase equipment that would control the humidity in their factory because their four-color printing process was sensitive to temperature and humidity. Fortuitously Willis Carrier was given the job and he designed a system that would control both variables. This innovation evolved into a patent (see Fig. 7.2) and several years later Carrier and other engineers left to form the Carrier Engineering Corporation that today is part of United Technologies Corporation. This move led to development of a new consumer technology that in turn changed the economic and demographic history of the Southern US, especially for the state of Florida. This story of young genius and entrepreneurship can be found repeated in many hard and soft media formats.

However, this story is much more complicated. Not to take anything away from Carrier's drive and accomplishments, we can't ignore that in 1902, he had just attended 4 years of an intensive engineering program in mechanical engineering with a concentration in electrical applications. The Director of the Sibley College at

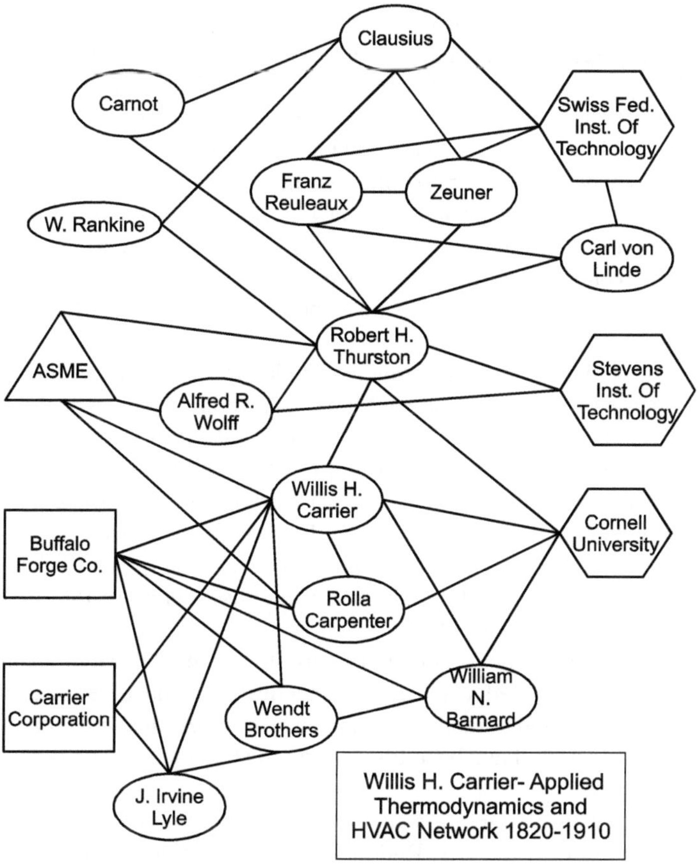

Fig. 7.3 Partial applied thermodynamics network centered around Willis H. Carrier and air conditioning and refrigeration machines

Cornell was Professor Robert H. Thurston, one of the most renowned authorities in applied thermodynamics. He was also one of the founding members as well as the first president of the American Society of Mechanical Engineers. Thurston had spent 14 years at Stevens Tech in NJ near New York City and helped establish an more academically and science-based engineering program there before coming upstate to Ithaca New York to take over the Cornell Sibley College in 1885.

All students at the Cornell Sibley College had to take Thurston's course on the steam engine that, according to the syllabus, contained a good dose of applied thermodynamics. Students also had to take a course in mechanical engineering laboratory taught by Professor Rolla C Carpenter that included experiments in applied thermodynamics including refrigeration. As it turned out, Carpenter was the author of two books on heating and ventilation of buildings that also contained material on cooling of buildings (Carpenter 1893).

Thus the farm boy Willis Carrier was immersed in a curriculum that taught new engineers to use science and math-based design methods. If there was no basic scientific theory to underpin the design, they were taught to construct experimental tables with data that would provide guidelines for design choices such as size, shape, materials etc. This is what Carrier did when he first came to Buffalo Forge. He convinced the owners, the Wendt brothers, that their designs were not based on scientific and experimental principles. Carrier was given charge of a department that would develop new data and charts to allow Buffalo Forge to produce efficient machines that would increase the Wendt brothers' profits. It was out of this new laboratory that Carrier developed his air conditioning system for Sackett-Wilhelm Lithography.

At the turn of the century, the principal nodes in the environmental control of building spaces were Carl von Linde in Munich, Alfred R. Wolff in the US and Willis Carrier. As we have seen in the early aviation network, Robert Thurston played the role of a secondary but enhancing node in the air conditioning machine network. Thurston had written a history of the steam engine that included a survey of the evolution of the subject of heat energy and thermodynamics. He had also written a popular book on the principles and applications of heat energy as well as translated the work of Sadi Carnot into English, one of the discoverers of the second law of thermodynamics (Thurston 1890).

One tool for applying thermodynamics was the use of a "thermodynamic cycle", and it was common knowledge at that time that design of machines to create cold or refrigeration could be understood in the same way one designed a steam engine. This knowledge would surely have been transmitted to students in Thurston's courses in the Sibley College.

Another player in refrigeration at the time was von Linde in Germany who had studied with Gustav Zeuner, Rudolf Clausius and Franz Reuleaux in ETH in Zurich from 1861–1864. Reuleaux was a secondary node in the internal combustion engine network as well as in the aviation network with his help of Lillienthal in Berlin. Reuleaux also helped von Linde get his first job. Rudolf Clausius along with William Rankine was one of the principal developers of thermodynamic theory. Gustav Zeuner went to the Technical University of Dresden and was one of the founders of technical thermodynamics. After Zurich, von Linde became a professor at the Technische Hochschule of Munich and developed machines to produce extreme cold which could liquefy air and other gases. In 1879, he left the university to start his own company, Linde AG, designing refrigeration machines for breweries. Von Linde's accomplishments cannot ignore his connections to the Technical University at Zurich, Zeuner and Clausius. In the same way Carrier's invention of a cooling and humidity control system cannot be viewed in isolation from his scientific and technical education at Cornell University.

To further complete the social network, shown in Fig. 7.3, we have the important node of Alfred R. Wolff who was a graduate of Stevens Institute of Technology when Thurston was a professor there. Wolff is credited with producing cooling systems for several important buildings in New York City before and around the time that

Carrier got the Sacket-Wilhelm Lithography job. Woolf also was a Charter member of the ASME of which Thurston was first president.

Another network link is that between Professor Carpenter and Buffalo Forge. Carpenter had written a book on ventilation in buildings around 1891 (Carpenter 1891, 1902). The back of one edition of the book contains advertisements and includes a full page ad from Buffalo Forge Co. Carpenter was also a consultant to many companies but we have not been able to make a stronger link between him and BFC. However a biography of Carrier, notes that BFC visited Ithaca to recruit Cornell students and that is how Carrier got a job. One may speculate that either Carpenter or Thurston knew one of the principles of Buffalo Forge.

Another Cornell node was William N. Barnard, an assistant professor during Carrier's studies at Cornell (Barnard et al. 1912, 1915). Barnard wrote a popular book in 1912, *Heat Power Engineering*, with two other faculty members. In the biography by Ingels (1952) mention is made of how Professor Barnard visited Buffalo Forge in 1902 to see how Cornell graduates such as Willis Carrier were doing. Thus Carrier continued to have contact with Cornell faculty after he went to Buffalo.

One of the hallmarks of the genius myth is the 'epiphany story' often told by the hero-inventor years after the event. In the example of Willis Carrier, this story appears in several biographical essays and books in which the key idea of dew point control came to him while he was on a foggy rail platform in Pittsburgh waiting for a train. This "genius myth" received official sanction in a millennium book on innovation in the twentieth century by the National Academy of Engineering in 2000. But the idea that Carrier invented the psychometric chart without other influences and single-handedly invented air-conditioning does not agree with the history of heat energy.

Oddly when the author was researching this network, a colleague at Cornell mentioned that he used this story in a motivational lecture to new students, to show that one of them might become a future Carrier and create of a new industry. So the genius myth still plays its role in educating the next generation of engineers and scientists.

As further evidence that air-conditioning technology followed a path similar to that of internal combustion engines and radios, we plot the growth in air-conditioning related events in Fig. 7.4a that shows again the exponential growth in a knowledge base. If we plot the same data on a logarithmic scale (Fig. 7.4b), we can see a straight-line region, indicating an exponential growth period from around 1840–1910. This milestone chart is based in part on a biography of Carrier edited by the Carrier Corp that he founded. This chart shows that Willis Carrier left Cornell in 1902 for his job with Buffalo Forge at a time when he was immersed in an avalanche of knowledge associated with the heating, cooling and ventilation of enclosed spaces using the ideas of thermodynamics.

One idea that emerges from this study is that patents and copyrights rarely slow the progress of human society in embracing some new idea or device. Sometimes it is not necessary for an outsider to gain direct knowledge of a trade secret. It is possible that the 'fact' that someone has accomplished some feat or created some new process is enough to motivate an outsider to try and duplicate, improve or optimize the invention or innovation. Imitation is an important feature of human learning.

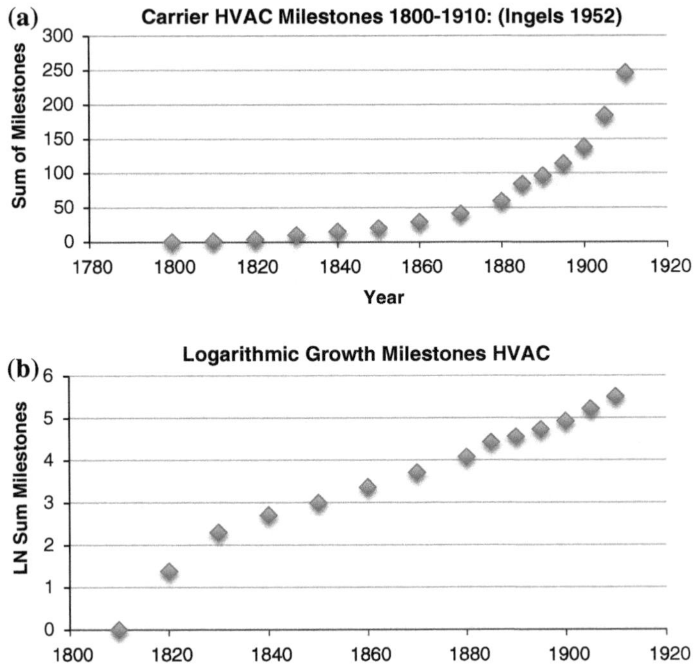

Fig. 7.4 **a** Graph of the sum of events associated with the thermodynamics and its application to heating, cooling and ventilation of enclosed spaces. **b** *Vertical scale* is the logarithm of the data in **a**. The *straight-line* region indicates an time period of exponential growth in knowledge in applied thermodynamics and HCAC (*Source*, Ingels 1952)

In the preceding chapters we have presented data on the exponential rise of innovation in a half dozen areas spurred on by the growth of social networks. The *rising tide of knowledge* in many technical and scientific areas is sometimes all that is necessary for someone to break out of the pack and promote new ideas, inventions and methodologies.

7.2 The End of the Genius-Inventor and the Rise of a Theory of Innovation

From the histories and data in Chaps. 1–6, we have seen common patterns in the evolution of invention and innovation in disparate areas, from steam engines to radios. Characteristic of all the technologies and scientific ideas we have studied is the simultaneous appearance of a network of individuals and institutions that grows exponentially as the milestones of innovation progress increase. We have used the analogy of an avalanche of sand in which small movements of particles continue to generate a distribution of smaller avalanches. This self-generation of motion has been called different names such as "*self-organized criticality*" or SOC by Per Bak (1996) and "strategic exchange in social systems" by G. D. Snooks (1998). Computational economists have used the term "agent-based modeling" or ABM as in e.g. Ball (2004) in his book *Critical Mass-How One Thing Leads to Another*.

A summary of the growth curves for eight of the different technologies and scientific theories discussed in this book are shown in Fig. 7.5. There is nothing to suggest that similar patterns of innovation advancement would not be seen in other technologies such as railroads, telephones, transistors or lasers. The only feature we can discern between steam engines and radios or between radio electronics and chaos theory is the decrease in the time scale to reach the peak of activity. If we plot the logarithm of the growth measure versus time, we obtain a measure of the growth time in years or decades to double the growth.

Estimates of the growth constants for the technologies shown in this book are shown in Table 7.1 and compared with that of modern innovation growth such as cell phone use or Facebook enrollees. There appear to be a continuous advancement in the pace of innovation from the eighteenth to the twenty-first centuries. What is remarkable about this data is the fact that qualitatively, innovation growth has remained similar across technologies and across the centuries. As forecast in Alvin Toeffler's 1970s book (Toeffler 1970), *Future Shock*, the pace of change has quickened but the processes of change seem to have similar structures governed by the growth of social networks in the histories of innovation and invention.

7.2.1 Summary of Innovation Growth Curves

The innovation growth curves in Fig. 7.5 can be plotted on a logarithmic vertical scale and the approximate slope of the resulting curve is the growth constant in units of 1/decade. The time to double the number of events or nodes in the network in given by $T = Ln2/GrowthConstant$. Table 7.1, shows the estimated growth constants for the various technologies and sciences. It is clear that the difference between technologies in the modern age and those a century ago is the significant decrease in the time to double innovation information or product penetration into the marketplace.

Another way to view this historical data is to plot the time to double the innovation growth measures versus historical time, as shown in Fig. 7.6. One can plot network nodes, products sold, technology/scientific events, etc. Fig. 7.6 must be taken as a very rough plot, but the message is that there had been a continuous decrease in innovation growth doubling time across technologies and spanning several centuries.

The validity of this conjecture will have to wait further historical studies into other technologies. One might speculate on whether this has the structure of a power law in the sense that city populations are ranked by a power law as well as scientific citations studied by De Solla-Price (1963). This conjecture is in the same spirit as the work of George Zipf (1949) in his attempt the bring mathematics to the subject of sociology. It also follows the suggestions of Per Bak (1996) in the possibility

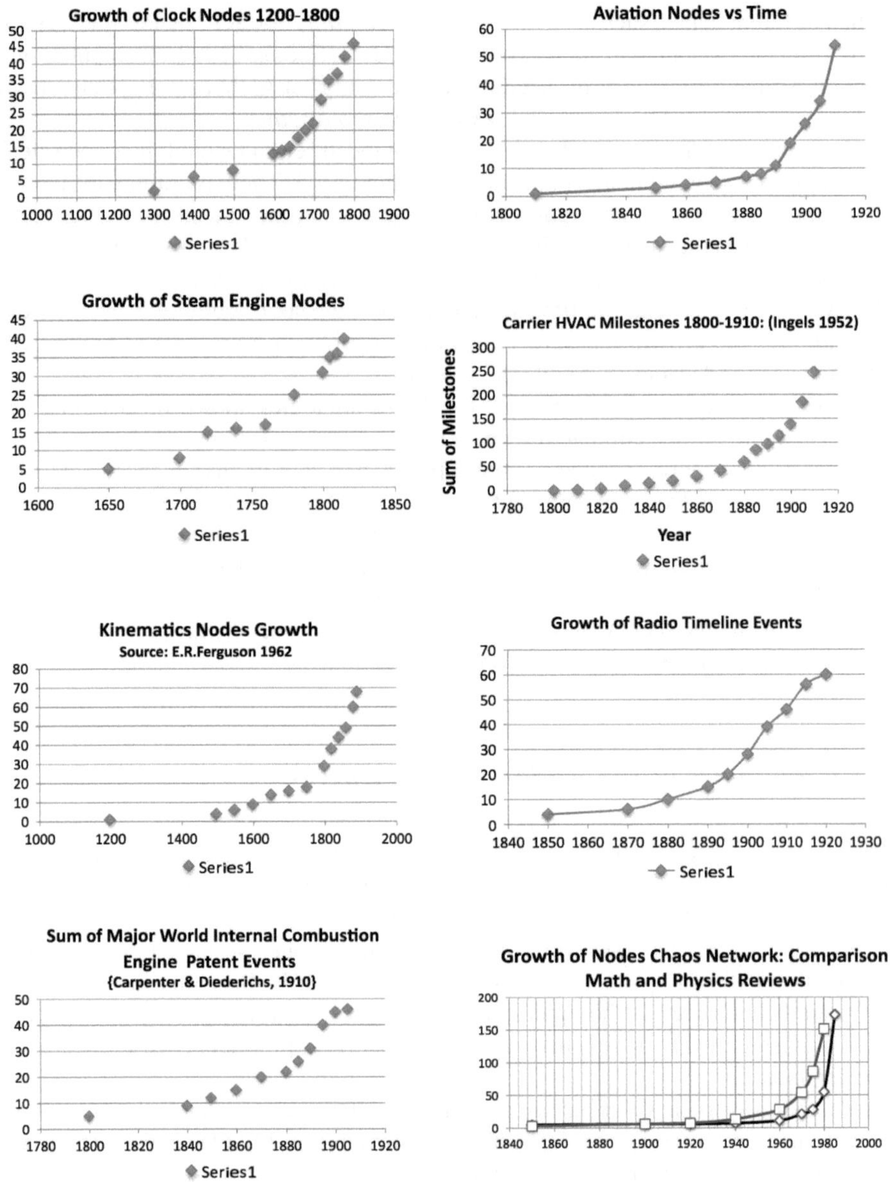

Fig. 7.5 Innovation growth *curves* for eight different technologies and sciences

of historical technical networks being governed by sand pile models and hence the natural occurrence of a power law behavior.

The reader might ask why the author did not follow up and see if other technologies would fit the model in Fig. 7.6. The simple answer is that the author did follow up

Table 7.1 Innovation exponential growth constants

Innovation type	Time frame	Exponential growth constant: 1/decade
Steam engine nodes	1650–1815	0.13
Steam engine growth (UK)	1710–1800	0.33
HVAC heat engineering nodes	1860–1910	0.50
Early aviation nodes	1880–1910	0.66
Internal combustion engine patents (US)	1880–1900	1.2
Wireless/Radio electronics nodes	1890–1920	0.65
Chaos theory nodes	1960–1985	1–1.8
Cell phone growth	1988–1998	3.2
Facebook users growth	2005–2010	9.2

since the original book was based on only four technical areas that grew into 10–12 technical and scientific studies. Although one can data-mine modern networks for thousands of data points as to usage, citations, network paths et cetera, historical networks require reading many books, reviews and other documents. Thus the task of exploring the power law conjecture of Fig. 7.6 must be seen as a work in progress.

7.2.1.1 Models for the Growth of Networks

The study of models for the growth of social networks is a large part of the study of modern network analysis. (See e.g. a recent textbook by Easley and Kleinberg 2010, ref in Chap. 1). There are problems of how a new idea propagates through an existing network, such as the adoption of some new crop seed in a community of farmers, or the creation of a larger network out of a set of clusters of smaller networks. Earlier we talked about the power-law property of networks in which a few nodes dominate the number of links as in the role of Google on the World-wide Web. Network textbooks call this the *rich-get-richer* effect. Attempts to understand this phenomenon began with Price's study of scientific paper citations in 1965 and was further developed by Barabasi and Albert in 1999 (see Newman (2003, ref in Chap. 1), for a discussion of network growth models). Price called this effect "*preferential attachment*" or the tendency for a new node to link up to the node in the network with the greatest number of links. This explanation may be good enough for an abstract mathematical model but historians might feel this is 'begging the question' as to why one technical network begins an exponential growth and another stalemates and declines.

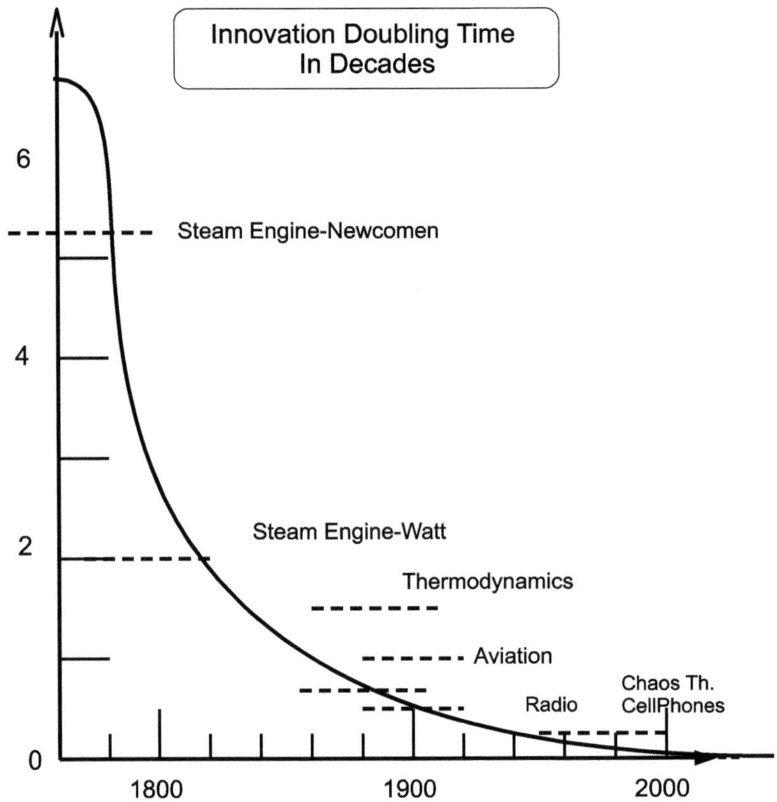

Fig. 7.6 Estimated decrease in the doubling time in decades for innovation across several technical and scientific fields based on the data in Fig. 7.5 and Table 7.1

7.3 Innovation Growth Models in Economics

Traditional economic theory was based on equilibrium models involving demand, supply and price as well as capital and labor. In the early twentieth century the Austrian-American economist Joseph Schumpeter [1883–1950] posited the idea that dynamic change was a better model and that one of the drivers of change was technological innovation. In Mokyr's book, *The Lever of Riches* (1990, ref in chap. 1), he uses the term *Schumpeterian growth*, defined as: *capitalist expansion driving from continuous, though fluctuating, technical change and innovation, financed by—credit.* In the wake of Schumpeter's ideas is a wealth of studies on innovation, especially the development of prediction models for innovation and diffusion of technical ideas and product into the marketplace. One recent collection of papers is *Economics of Innovation*, edited by Hall and Rosenberg (2010), with 29 articles in this two-volume work. A perusal of a university library section on economics of innovation will reveal dozens of works and hundreds of research papers relating to technological innovation, invention, research and development (R&D) and diffusion of products.

What is the difference between economic studies of innovation and the history-based network theory of innovation discussed in this book? Economic studies generally focus on a shorter time scale: for example the time for corporate investment in innovation (R&D) to reap profits, or the time scale for a product to reach saturation. Historical studies of inventions often look at time scales of decades and centuries. A second difference between history-based and economic-based innovation studies is the nature of the questions asked. In history of technology we usually ask 'who, when and where' and in economics one asks 'how much'. Economic analysis often uses statistical data and probability models. In history of technology we are searching for specific connections, links, between each inventor node and other nodes in the network. There are exceptions of course. In a recent economic study from Germany, *Networks in the Innovation Process* (Graf 2006), the author draws network diagrams between industrial companies and research institutions around the German city of Jena [see Figs. 6.2 and 6.3, in Graf (2006)].

It is interesting to speculate on why economists seem to have a greater desire to understand the process of innovation than the engineering community who are often the genesis of technological invention. Inventors and innovators are those who look to the future and have traditionally shown little interest in how they create.

7.4 Fading Memory in Historical Innovation Networks

We have assumed that nodes promulgate across generations in the networks discussed in this book. This is tantamount to saying that knowledge generated by earlier nodes is of value to later persons and institutional nodes in the network. For example, Lenoir's experience in 1865 with a stationary internal combustion engine was relevant to the Otto-Langen engine of 1876, but was not likely useful for the design of the Daimler lightweight engine of 1896 in dirigibles. As the internal combustion engine network was growing, there were earlier nodes and links that were disappearing. We call this phenomenon *fading memory*. Like an old man who knows the latest football players but forgets the names of his old friends, innovation networks have both growing and dying branches. Another way of saying this is that the information commons for any given field can shrink as well as expand.

As an anecdote consider the field of electric batteries, the so-called voltaic cells of the nineteenth century. This is part of the field of electro-chemistry. Electric batteries were of importance to both chemistry and physics laboratories as well as the telegraph and telephone businesses of the late nineteenth century. To test our thesis that networks have fading memory we examined two popular textbooks on physics from 1899 and 1955. The French physics text by Professor Ganot was originally published in 1850 and translated into English by E. Atkinson through 15 editions. *University Physics*, by F.W. Sears and M.W. Zemansky, published in 1949 and 1955, was a popular freshman text of its day. Each book has a section on electricity and electro-chemistry and batteries.

In Ganot (1899, Sect. 822), Voltaic cells are described in which two metals in a conducting liquid are used to generate an electric voltage. The voltaic battery is a series of voltaic cells based on copper and zinc in "acidulated water or brine". In the 1899 edition a set of improved batteries is described ; Wollaston's battery (Sect. 826), Daniell's battery (Sect. 829), Grove's battery (Sect. 830), Bunsen's battery (Sect. 831), Smee's battery (Sect. 832) and a number of "recent batteries by Marie Davy and Meidinger" (used for telegraph systems). These batteries used copper, zinc, carbon and platinum. No mention was made of lead-acid batteries used in today's automobiles.

In Sears and Zemansky (1955), there is also a section on galvanic cells or electrochemical batteries. Of the list of batteries in Ganot (1899), only the Daniell cell with zinc and copper is described in Sears and Zemansky. The lead-storage battery is described in detail (Sect. 30-11) as well as a Weston standard cell using mercury and cadmium from the Weston Electrical Instrument (Sect. 30-12). The names of Bunsen, Wollaston, Smee stc have disappeared from the electric battery information commons of 1955.

Another example using the same two physics texts of 1899 and 1955 is from the field of thermodynamics, relevant to the earlier case study in this chapter on air conditioning. Ganot (1899) has a 220 page section *On Heat* (Book VI) and Sears and Zemansky (1955) has a section "*Heat*" comprising Chaps. 15–20.

In a section on thermometers, Ganot lists mercury, alcohol and differential thermometers as well as Brequet's metallic thermometer and Sixes' thermometer. In Sears and Zemansky they describe 'liquid-in-glass, thermometers and mercury-in-glass thermometers as well as thermocouples, optical pyrometers and a constant-volume gas thermometer. Brequet and Sixes have disappeared in the thermometer network of 1955.

Yet another example from these two books are descriptions of calorimeters and the measurement of heat. Ganot (1899) lists the names of Favre, De la Roche, Regnault, Black and Bunsen calorimeters. These names have disappeared in the 1955 physics book of Sears and Zemansky. In Ganot there is a description of a steam engine or "Cornish engine" and no mention of the second law of thermodynamics. In 1955, Sears and Zemansky reduced steam engines to the Rankine cycle and internal combustion engines to Otto and Diesel cycles without any technical details of the machines' construction.

Perhaps we have belabored the point with the minutia of physics texts, but the message is that innovation networks die in parts as well as grow. When the old branches die faster than new nodes are added, the network dies completely as one can easily find in the debris of technology and science as in the fields of vacuum tube electronics and wave propagation in the ether. In this book we have focused only on growing innovation networks and leave the withering and death of such networks to another book.

7.5 The Role of the Genius-Hero in Innovation Networks

In our construction of invention and innovation networks in this book we have come across the following genius-innovator and hero-inventors:

James Watt: Steam engine;

Leonardo da Vinci: Construction of machines;

Franz Reuleaux: Kinematics of mechanisms;

John Harrison: Precision clocks;

Nikolas Otto: Four-cycle internal combustion engine;

Henry Ford: mass production of automobiles;

Montgolfier brothers: early balloon flight;

Wright brothers: controlled flight of airplane;

Gugliermo Marconi: Wireless telegraph;

Edwin Armstrong: modern radio circuits;

Balthazar van der Pol: mathematical analysis of electronic circuits;

Edward Lorenz: Modeling chaotic behavior of weather;

Willis H. Carrier: modern air conditioning technology.

What characterizes the hero-inventor/genius in the network study of innovation? In the traditional study of technology, often the hero is created by the person or the company that he/she founded. Sometimes the image or 'branding' is created by the media.

7.5.1 The Genius Theory of Innovation

In this book we have examined the networks attendant to these popular heroes of technology. Innovators and inventors have often been lionized by biographers and historians asleading the way to a new world of technology. We have tried however to demonstrate that each of these men were surrounded by a cadre of equals, supporters and assistants who contributed essential ideas and talents to their success. It is likely that a detailed investigation of other paragons of invention such as Morse, Siemens, Edison, Bell, Madame Curie or even Steve Jobs would reveal similar results. Does this mean that the singular hero-inventor is dead?

In constructing social historical networks, we have found that the hero-inventor is often a node that generates a large number of incoming and outgoing links. It is in this sense that he or she is crowned a pioneer or leader in the field. It is not the one who is first with an innovation who is awarded the mantle of the hero-inventor, but the one who brings together both new and old knowledge, (generating incoming links) and inspires others to copy and improve his or her invention and innovation (generating outgoing links).

Thus we are not advocating a purely collective theory of innovation. The demise of the Soviet system provides evidence that discouragement of the creative talents of the individual and submission to the collective plan was not a good strategy for Soviet society. However the dual nature of the innovator and the social network is a dynamic we do not fully understand and historical analysis of earlier technologies may provide clues as to how it worked in the past and how it works today.

7.5.2 Epiphany Stories of the 'Genius-Hero'?

We have downplayed the importance of the epiphany story of the hero-inventor in order to argue for the role of the network within which the inventor was living and working. However the epiphany story is important in society as a motivational tool and a cultural marker of the importance of creating new ideas and products in the modern world. Our point however is that the recitation of these tales is no substitute for historical accuracy. Most modern corporations recognize that innovation success requires both creative individuals as well as a network of different talents. How to manage the tensions between the team and the genius in an organization is what business schools are for. The message in this book however is that the innovation network extends far beyond the corporate corridors. And as many organizations have discovered, so-called trade secrets sometimes are already imbedded in the knowledge commons beyond the corporation.

7.6 Role of the Media and the Knowledge Commons

In the construction of historical innovation networks we have included magazines and journals as nodes. In several examples we have seen the importance of books, magazines and scientific publications in providing links for the innovation network. The importance of books is seen in the 'theatre of machines' works published between the fifteenth and eighteenth centuries (see Chap. 2). In the nineteenth century periodicals such as *Scientific American*, *Engineering* (London), *Cassier's Magazine* and *American Artisan* became nodes for the internal combustion engine and other technologies. They often presented extensive reports on international expositions where new machines were on display to generate sales and extend markets. Often new technologies create their own media as in the rise of the automobile with specialized magazines such as *Horseless Carriage*, or in aviation, *Aeronautical Journal* and in radio, *Electrical World*. Technically interested people regularly read these magazines in the nineteenth and twentieth centuries as we today scan the Internet for new ideas.

7.6.1 The Knowledge Commons

Though not treated as a separate node, in every field there accumulates a set of common knowledge as the field matures. This knowledge could take the form of general principles, empirical data, rules of thumb, misconceptions, recipes, formulas, topologies and geometries. General principles include Newton's laws of mechanics, data such as the strength of certain steels, recipes for the mixture ratios of copper and zinc, topologies such as the connections in an electrical circuit. This knowledge is passed around through handbooks, journals and magazines, expositions, artisan trial and error experience, technical meetings and personal correspondence. In the author's library there is an 1890 *Engineer's Sketch-Book* of mechanical movements, published by an engineer, Thomas W. Barber in London. This little book has 1936 sketches of mechanisms and machine details. Similar books appeared in the early twentieth century for electronic circuits. In the 1970's the Soviet Union published an English translation of Artobolevsky's handbook of kinematic mechanisms with over 5000 mechanisms.

The author has a 1930, 3rd edition of *Mark's Handbook* (Marks 1930), a mechanical engineering book first published in 1916 by Lionel Marks, a professor at Harvard and another graduate from the Cornell Sibley College of Mechanical Engineering at the time of Robert Thurston. In the 1930 edition, some of the contributors include, Willis Carrier, J.P. Den Hartog and Steven Timoshenko (see Chap. 6). Mark's Handbook is still published today. These treasuries of technical and scientific wisdom, published in many languages, are rarely mentioned or acknowledged by historians as having any role in the creation of technology. Yet as of 1930, there were 118,500 copies of *Mark's Handbook* sold.

In the study of political science and law there is a wider concept of '*the commons*' such as free public access to drinking water, parks, roads and school knowledge. Recently in a book *Common as Air*, writer Hyde (2010) discusses so-called 'intellectual property' and to what extend literary text, art and music belongs to the public commons. In this chapter we take a more limited definition of the *knowledge commons*. A pedestrian example is the knowledge one can find by spending time in an automotive garage while auto mechanics are fixing cars. The untrained person could very quickly discover what were the basic components of a car, where they were placed in different models and makes, what systems needed repair most often, what lubricants and fuels were important in a modern car and how much things cost if they are to be replaced.

In the later years of the Cold War, the author was carrying out research on magnetic levitation of trains and had attended an international meeting in Japan at which some Soviet engineers were present. After the meeting an agent from an intelligence agency visited the author and asked if he had seen any important ideas at the meeting from the Soviet delegation. The agent was surprised when he learned he could have saved a trip since all the papers were published in a proceeding and were available to the public.

The knowledge commons most likely helped Charlie Taylor, the Wright brothers technician, build a lightweight internal combustion engine (12 hp, 220 pounds) for their 1903 flight (Crouch 1989). Many college glider clubs in the first decades of the twentieth were able to build human carrying gliders with shared technical details and published results from popular technical magazines. The knowledge commons was also a factor in creating the first electronic vacuum tube circuits that inspired the early amateur radio clubs.

7.7 Innovation Networks and the 'Paradigm Shift'

One reviewer of a preprint of this monograph asked if the network view of innovation had any relevance to Thomas Kuhn's famous book, *The Structure of Scientific Revolutions* (Kuhn 1962). Kuhn [1922–1996] was a physicist and philosopher of science. He described a process whereby 'normal' scientific thinking in a research community using classical problems or 'paradigms', eventually leads to conflicts with accepted ideas. This conflict produces a new, radical paradigm that triggers a scientific revolution. Two often cited examples are Einstein's theory of relativity and quantum physics. Kuhn's theory suggests that science proceeds in discontinuities, so-called "paradigm shifts" that appears to be at odds with the evolution model of science and technology put forward in this monograph.

Applied to technological innovation, the development of the internal combustion engine created a dramatic paradigm shift in the way machines were designed and enabled the creation of human carrying flying machines. The vacuum tube played a similar role in wireless communication and radio. In both cases the 'paradigm shift' was preceded by an exponential expansion of a social network. In our example of a network in scientific theory, the new ideas of nonlinear chaotic dynamics destroyed the 'normal' paradigm of a predictable Newtonian macro-science without quantum ideas. Here again the discoveries of Edward Lorenz in 1963, were proceeded by an earlier community of mathematicians, scientists and theoretical engineers. On the face of it, Kuhn's *paradigm shift theory* does not seem at odds with the idea of a social innovation network. His theory seems to embrace a community of scientists who both adopt a classic concept and eventually reject it.

The question of the role of the community or society in advancing science is a hotly debated topic in the philosophy of science and sociology. We have cited the work of Derek De Solla-Price (1963) and Collins (1998) as two thinkers who examined the role of community and networks in science. In our work we have introduced the idea of a "knowledge commons" that contributes to the social network in producing innovations. However in advancing the ideas of social networks and knowledge commons we do not mean to suggest that scientific ideas depend on or are relative to community values or beliefs, as a few modern writers have put forward. In technology, there is nothing tentative about the lift on a curved airfoil or the resonance in a regenerative vacuum tube circuit. Aircraft are designed by engineers who come from dozens of different cultures, and airplanes don't fall from the sky when they fly over a cultural region that does not believe in modern physics. We are only discussing the *process* of innovation through networks and do not venture into the realm of cognitive science and creativity in the individual.

Finally as to the singular nature of Thomas Kuhn's theories of scientific process, the social network apparently played a part in his writings. An earlier scholar, Michael Polanyi [1891–1976] from the University of Manchester, is sometimes described as having influenced Kuhn's *Scientific Revolutions* book.

7.8 A New Approach to the Study of Innovation History

In this book we have posited an alternative approach to the historical narrative of invention and innovation based on networks and historical data modeling. Graphics, statistics and mathematics can't replace the traditional narrative history but are intended to complement the classical study of history of technology and science. For reference we summarize the methodology presented in this book.

Source Materials

The use of traditional biographical and historical studies of inventors and comprehensive treatises of technologies are of great value in constructing networks especially establishing the links between nodes.

Original documents from archives of inventors and scientists are very useful in establishing links between nodes especially those that are overlooked in biographies.

Web-based material can save time, especially Wikipedia, but should be heavily supplemented with traditional textural and archival material. Despite the promise of web-based libraries in the future, most textural material has not been scanned by Google, especially obscure technical treatises and documentation.

Graphical Analysis

Traditional timeline plots are useful in identifying principal nodes.

Star network diagrams centered around principal nodes are the next step.

The complex integration of star networks into a comprehensive network for an innovation field is a difficult task. It involves some aesthetics to make it readable. The lack of color in this book limited the information categories that could be used such as country of origin or principal or secondary nodes or enabling nodes. The vertical orientation of traditional books meant that time had to run from top to bottom whereas in an art setting, one would move time horizontally, typically from left to right. In constructing the network online, one can also use CAD tools to construct 3D networks as is done for complex molecules such as DNA.

Integration of separate networks from different fields, such as electronics and nonlinear dynamics in Chap. 6 is a difficult task to show in a book form and might best be done in a web-based CAD environment.

Network Statistical Analysis

We have tried to use only the most elementary mathematical ideas from network theory and related fields. We introduced the *integrated knowledge growth function*, as a measure of the maturity of an innovation field and showed that some fields grow linearly and others have an exponential growth.

Different data bases have been used such as growth of product market penetration, number of patents, numbers of journals and publications and in the science world one can use numbers of citations as did de Solla Price in earlier studies of science.

The modern use of link-node statistics might be useful to establish hub nodes and small worlds and clusters of nodes. The problem with these tools is that for modern social networks such as the World-Wide Web or Facebook, statisticians have millions of data points to use whereas in the innovation history networks, studied in this book, one has only a few dozen nodes and perhaps a few hundred links. Nonetheless, humans are always making estimates from small samples. Here one must use judgment when interpreting small sample data.

One of the tools illustrated in Chaps. 1, 4, was the *influence matrix*, used for the steam engine and aviation networks. This technique could be further explored and perhaps codified to automatically calculate link-node statistics, shortest paths distributions, growth of links and nodes in time.

Mathematical Models for Innovation Growth

This is the weakest part of this book and only a few suggestions have been made in Chap. 1. Certainly the model types used in ecology and population dynamics might be used. Also chemical kinetics models for reacting atoms and molecules might suggest a path to predicting the growth of innovation and invention networks and links. Economic theorists have posited numerous models for innovation growth that may be relevant to historical studies.

Recent ideas from Complexity Theory such as *self-organized criticality*, are fascinating and may provide clues as to how human societies create new ideas, products and processes. What is missing is the mechanism for the micro and macro avalanches, or the slow and sudden growth modes of the links and nodes in the innovation network. Modern network analysis have also provided models for network growth based on probability measures. Economic models have sometimes used game theory approaches. But perhaps the historian of the future will have to look at cognitive science, to look inside the human brain, to see what motivates humans to create and embrace new ideas and inventions.

7.9 Hallmarks of Innovation

What is necessary for a culture to produce new ideas and technologies? In this book we have made the case for a societal role in innovation through historical networks. Historical and economic evidence has shown that there are a number of other societal conditions that must be present to incubate new ideas and technologies. Some of these conditions include the following.

(1) The society must have a tradition of creating new ideas and machines;
(2) There must exist a cadre of artisans and craftspeople with technical skills;
(3) There must exist a supply of capital to invest in new technology;
(4) There must exist a spirit of progress, that humankind is meant to improve and change its environment;
(5) The society must have a governmental structure that encourages change, new ideas and security for investors;
(6) There must exist individuals with a vision and motivation to change the status quo.

In the so-called middle ages, before the Renaissance, there was a growth of cities and guilds that produced skilled craftspeople. The Scholastics in the Church schools developed ideas of reason and progress as part of 'God's plan' for humanity. The merchant class developed and accumulated capital and a need to invest that capital. Also during the pre-Renaissance we saw trades emerge to create fantastic cathedrals and castles that required skills to organize workers and develop long term goals. Out of this milieu evolved the early machine age of Francesco di Giorgio and Leonardo da Vinci and the inexorable march to the industrial revolution of the nineteenth century age of machines and the twentieth century the age of electronics and information.

Yet as historians have painstakingly gathered evidence for this evolutionary model of scientific and technical progress, biographers have steadfastly maintained the myth of the genius inventor and innovator. The tension that arose between techno-historians and romantics in the eighteenth and nineteenth century continues to exist today.

Modern historians have tried to describe the last quarter century as a *Knowledge Revolution* in contrast to the Industrial Revolution based on machines. In pursuing the networks behind inventions in this book, we have used the idea of a *knowledge commons* active both before and during the Industrial Revolution, that helped evolve and spread the innovation concept and its attendant social networks. Humans have always organized themselves into networks to preserve and protect useful knowledge and the only difference between the present and past history is the amount of knowledge and the time scale in which it evolves.

References

Bak P (1996) How nature works: the science of self-organized criticality. Springer, New York

Ball P (2004) Critical mass: how one thing leads to another. Farrar, Straus and Giroux, New York
Barnard WN, Ellenwood FO, Hirshfeld CF (1912, 1915) Heat-power engineering, 3rd edn. Wiley, New York
Buchanan M (2000) Ubiquity: the science of history. Weidenfeld and Nicolson, London
Carpenter RC (1891, 1902) Heating and ventilating buildings: a manual for heating engineers and architects. Wiley, New York
Carpenter RC (1893) Experimental engineering. Wiley, New York
Collins R (1998) The sociology of philosophies. Harvard University Press, Massachusetts
Crouch T (1989) The Bishop's boys. W. W. Norton, New York
De Solla-Price DJ (1963) Little science. Big Science Columbia University Press, New York
Ganot (1899) Elements de Physique, (Elementary Treatise on Physics, Translated by E. Atkinson), 15th edn. William Wood, New York
Graf H (2006) Networks in the innovation process. E.Elgar Publ Northampton, MA
Hyde L (2010) Common as air. Farrar, Struas and Giroux, New York
Ingels M (1952) Willis haviland carrier: father of air conditioning. Country Life Press, New York
Johnson S (2010) Where good ideas come from: the natural history of innovation. Penguin Books, New York
Kuhn TS (1962) The structure of scientific revolutions. University of Chicago Press, Chicago
Lock CGW (1882) Spon's encyclopedia of the industrial arts. E & F N Spon, London
Marks LS (1930) Mechanical engineers' handbook, 3rd edn. McGraw-Hill Book, New York
Sears FW, Zemansky MW (1955) University physics, 2nd edn. Addison-Wesley, Massachusetts
Snooks GD (1998) The laws of history. Routledge, London
Thurston RH (1890) Heat as a form of energy. Houghton Mifflin, Boston
Toeffler A (1970) Future shock. Random House, New York
Wampler C (1989) Dr. Willis H, Carrier: father of air conditioning. Newcomen Society of England, American Branch, New York
Zipf GK (1949) Human behavior and the principle of least effort. Addison-Wesley Press, Cambridge

Subject Index

F. C. Moon, *Social Networks in the History of Innovation and Invention*,
History of Mechanism and Machine Science 22, DOI: 10.1007/978-94-007-7528-2,
© Springer Science+Business Media Dordrecht 2014

Author Index

F. C. Moon, *Social Networks in the History of Innovation and Invention*,
History of Mechanism and Machine Science 22, DOI: 10.1007/978-94-007-7528-2,

© Springer Science+Business Media Dordrecht 2014

MIX
Papier aus verantwortungsvollen Quellen
Paper from responsible sources
FSC® C105338

If you have any concerns about our products,
you can contact us on
ProductSafety@springernature.com

In case Publisher is established outside the EU,
the EU authorized representative is:
Springer Nature Customer Service Center GmbH
Europaplatz 3, 69115 Heidelberg, Germany

Printed by Libri Plureos GmbH
in Hamburg, Germany